国外优秀数学著作
原　版　系　列

与广义积分变换有关的分数次演算
——对分数次演算的研究

Fractional Calculus Associated With Generalized Integral Transformation
—A Study of Fractional Calculus

（英文）

[印] 哈门德拉·库马尔·曼迪亚（Harmendra Kumar Mandia）
[印] 亚什万特·辛格（Yashwant Singh）　著

哈尔滨工业大学出版社
HARBIN INSTITUTE OF TECHNOLOGY PRESS

黑版贸审字 08－2019－180 号

图书在版编目(CIP)数据

与广义积分变换有关的分数次演算：对分数次演算的研究＝Fractional Calculus Associated With Generalized Integral Transformation：A Study of Fractional Calculus：英文/(印)哈门德拉·库马尔·曼迪亚(Harmendra Kumar Mandia)，(印)亚什万特·辛格(Yashwant Singh)著. —哈尔滨：哈尔滨工业大学出版社，2023.1
ISBN 978-7-5767-0602-4

Ⅰ.①与… Ⅱ.①哈…②亚… Ⅲ.①分数次积分－英文 Ⅳ.①O172.2

中国国家版本馆 CIP 数据核字(2023)第 031372 号

YU GUANGYI JIFEN BIANHUAN YOUGUAN DE FENSHUCI YANSUAN：DUI FENSHUCI YANSUAN DE YANJIU

策划编辑 刘培杰 杜莹雪
责任编辑 刘家琳
封面设计 孙茵艾
出版发行 哈尔滨工业大学出版社
社　　址 哈尔滨市南岗区复华四道街 10 号 邮编 150006
传　　真 0451－86414749
网　　址 http://hitpress.hit.edu.cn
印　　刷 黑龙江艺德印刷有限责任公司
开　　本 880 mm×1 230 mm 1/32 印张 8.5 字数 250 千字
版　　次 2023 年 1 月第 1 版 2023 年 1 月第 1 次印刷
书　　号 ISBN 978-7-5767-0602-4
定　　价 48.00 元

DECLARATION BY THE CANDIDATE

I declare that thesis entitled '**A study of fractional calculus associated with generalized integral transformation with applications**' Is my own work conducted under the supervision of **Dr. Yashwant singh** at Shri jagdish prasad jhabarmal Tibrewala University Chudela, Jhunjhunu (Rajasthan), Approved by research Degree Committee. I have put in more than 200 days of attendance with the supervisor at the centre.

I further declare that to the best of my knowledge the thesis does not contain any part of any work which has been submitted for award of any degree either in this University or any other university/ deemed university without proper citation.

Signature of Supervisor
(With stamp)

Signature of candidate
Harmendra Kumar Mandia

Signature of the Head/Principal
(With stamp)

CERTIFICATE OF SUPERVISOR

This is to certify that work entitled "A study of fractional calculus associated with generalized integral transformation with applications". Is a piece of research work done by Shri **HARMENDRA KUMAR MANDIA** Under my supervision for the degree of Doctor of philosophy in Mathematics of JJT University, Jhunjhunu, Rajasthan, India. That the candidate has put attendance of more than 200 days with me.

To the best of my knowledge and belief the thesis

I. Embodies the work of candidate himself

II. Has duly been completed

III. Fulfills the requirement of ordinance related to Ph.D. degree of the University and

IV. Is up to the standard both in respect of content and language for being referred to the examiner.

Signature of the Supervisor

(With stamp)

ACKNOWLEDGMENTS

I wish to express my heartfelt indebtedness to my supervisor Dr. Yashwant Singh, Lecturer, Department of Mathematics, Seth Motilal (P.G.) College Jhunjhunu,Rajasthan for his constant inspiration supervision and valuable guidance, liberal attitude and encouragement in making this endeavor a success. Without his meritorious discussions and fruitful criticisms it would not have been possible for me to accomplish this arduous task. Work with him was a pleasurable, enriching and memorable experience of my life and I take this opportunity to pay my sincere thanks and best regards to him

Thanks are due to Dr. Satyaveer singh and Dr. Atul Garg for giving valuable suggestions and taking interest in my research work.I am also grateful to Shri P.S. Sundria who has been a constant source of inspiration and encouragement.

I owe my special thanks to my wife Anurita who always stood by me with her moral and emotional support. A small little thank is reserved for my daughter Chitransha and my son Lalitaya for letting me compile this thesis even at night sometimes. My parents and Brother Rajendra kumar, Mrs. Suman and little child Mimansha also gratitude for encouragement and providing family atmosphere for me through the conduct of this research work. At this precious moment we can not forget to thanks our extreme friend Mr. Dinesh kumar for the computer support used in carrying out this work. Thanks are also due to Mr K.R. poonia, Mr. R.S. poonia, Mr. Anil Baloda, Mr. Virendra Singh for their kind help, suggestion and assistance provided throughout the work.

HARMENDRA KUMAR MANDIA

Department of Mathematics

Shri jagdish prasad jhabarmalTibrewalaUniversity,

Chudela, Jhunjhunu-333001Rajasthan (India)

CONTENTS

ABSTRACT

The present thesis entitled "A study of fractional calculus associated with generalized integral transformation with applications" is the outcome of the research work carried out by me since July 2010, under the able guidance and kind supervision of Dr. Yashwant Singh, Lecturer Department of mathematics, Faculty of science, Seth Motilal (P.G.) College, Jhunjhunu, Rajasthan.

The following research papers form the basis of the thesis:

The relation between double laplace transform and double hankel transform with applications.

Research and Review: J. Physics, Vol.1,No.1,(2012),24-30

(In collaboration with Dr. Yashwant Singh)

Relationship between double laplace transform and double mellin transform in terms of generalized hypergeometric function with applications.

Int. J. Scientific and Engineering Research, Vol. 3, No. 5,May- (2012)

(In collaboration with Dr. Yashwant Singh)

On the two dimensional weyl fractional calculus associated with whittaker transform.

Int. J. Computer Science and Emerging Technologies, Vol. 2, No. 5 (2011), 553-556.

(In collaboration with Dr. Yashwant Singh)

On some kober fractional q-integral operator of the basic analogue of the $\overline{H}$- function.

Int. j. Theoretical and Applied Physics, Vol. 1 No. 1 (2011), 53-62.

(In collaboration with Dr. Yashwant Singh)

On some multiplication formulae for generalized hypergeometric functions.

Int.J.Mathematical Science and Engineering Application, Vol. 6,No. 11, (2012), 39-45.

(In collaboration with Dr. Yashwant Singh)

A study of some transformation formulas involving I -function.

The Mathematical Education, Vol.157,No.2,(2013),Accepted for publication

(In collaboration with Dr. Yashwant Singh)

A study of unified theorems involving the laplace transform with application.

Int.J.Research and Reviews in Computer Science, Vol. 2,No. 6 (2011),1319-1322.

(In collaboration with Dr. Yashwant Singh)

Convolution integral equation with kernel as a generalized hypergeometric function and H -function of two variables.

Canadian J. Science and Engineering Mathematics,Vol. 3,No.1(2012),43-47.

(In collaboration with Dr. Yashwant Singh)

On composition of generalized fractional integrals involving product of generalized hypergeometric functions.

Int.J.Physical science , vol.23,no.3(2011), 727-737.

(In collaboration with Dr. Yashwant Singh)

On some integrals and fourier series involving generalized hypergeometric functions.

Int. J.Theoretical And Applied Physics, Vol.2,N0.1(2012),1-10.

(In collaboration with Dr. Yashwant Singh)

A unified study of astrophysical thermonuclear functions for boltzmann-gibbs statistics and tsallis statistics and $\overline{H}$ -function.

Int.J.Comp.Sci.Emerging Tech., Vol.2,No.6,(2011),348-353.

(In collaboration with Dr. Yashwant Singh)

The thesis consists of the following seven chapters:

At the end of the thesis an extensive bibliography have been given.

In chapter 1 we present a brief survey of the Fractional calculus,fractional integration operatore and generalized hypergeometric functions. Definations of I-, G-, H-, A-functions of one and more variables are given. Some of which will be employed in presenting the result of the subsequent results.

In chapter 2, section 1 we establish a relation between the double Laplace transform and the double Hankel transform.

following formula is required in the proof:

$$\int_0^\infty\int_0^\infty x^{s-1}y^{t-1}H\left[ax^\lambda, by^\mu\right]dxdy=\frac{a^{-s/\lambda}b^{-t/\mu}}{\lambda\mu}\phi\left(-\frac{s}{\lambda},-\frac{t}{\mu}\right)\theta_2\left(-\frac{s}{\lambda}\right)\theta_3\left(-\frac{t}{\mu}\right)$$

The H-function of two variables essentially to Srivastava and Panda is defined and represented as:

$$H[x,y]=H\left[\begin{smallmatrix}x\\y\end{smallmatrix}\right]=H^{0,n_1:m_2,n_2:m_3,n_3}_{p_1,q_1:p_2,q_2:p_3,q_3}\left[\begin{matrix}x\\y\end{matrix}\middle|\begin{matrix}(a_j;\alpha_j,A_j)_{1,p_1}:(c_j,\gamma_j)_{1,p_2},(e_j,E_j)_{1,p_3}\\(b_j;\beta_j,B_j)_{1,q_1}:(d_j,\delta_j)_{1,q_2},(f_j,F_j)_{1,q_3}\end{matrix}\right]$$

$$=-\frac{1}{4\pi^2}\int_{L_1}\int_{L_2}\phi(\xi,\eta)\theta_2(\xi)\theta_3(\eta)x^\xi y^\eta d\xi\, d\eta$$

In section 2 we establish a relation between the double Laplace transform and the double Mellin transform. A double Laplace-Mellin transform of the product of H-functions of one and two variables is then obtained.

If $F(p_1,p_2)$ is the laplace transform and $M(p_1,p_2)$ is the Mellin transform of $f(t_1,t_2)$, then

$$F(p_1,p_2)=\sum_{s_1=0}^\infty\sum_{s_2=0}^\infty\frac{(-p_1)^{s_1}}{s_1\,!}\frac{(-p_2)^{s_2}}{s_2\,!}M(s_1+1,s_2+1)$$

In chapter 3, Section 1 we derive a new theorem concerning the Whittaker transform of two variables. The result is derived by the application of two dimensional Erdelyi-Kober operators of Weyl type. Some known and new special cases are also given in the end.

Let

$$g(p,q) = W_{\lambda_1,\mu_1}^{\lambda,\mu}[F(x,y);\rho,\sigma,p,q] = \int_b^\infty\int_d^\infty (px)^{\rho-1}(qy)^{\sigma-1}$$

$$\exp\left(\frac{1}{2}px + \frac{1}{2}qy\right)W_{\lambda,\mu}(px)W_{\lambda_1,\mu_1}(qy)F(x,y)dxdy$$

Be the two-dimensional Whittaker transform, for $\alpha > 0, \beta > 0$, the following result holds:

$$K_p^{\eta,\alpha}K_q^{\delta,\beta}[g(p,q)] = G_{\lambda_1,\beta,\delta,\mu_1}^{\lambda,\alpha,\eta,\mu}[F(x,y);\rho,\sigma,p,q],$$

The two-dimensional Whittaker transform reduces to a two-dimensional Laplace transform and consequently, we have a result given by Saxena et. al.

In section 2 we derive an expansion formulae for a basic analogue $\overline{H}$-function have been derived by the applications of the q-Leibniz rule for the type q-derivatives of a product of two functions. Expansion formulae involving a basic analogue of Fox's H-function, Meijer's G-function and MacRobert's E-function have been derived as special cases of the main results.

The main results to be established are as under:

$$\overline{H}_{p+1,q+1}^{m+1,n}\left[\rho\left(zq^{\mu}\right)^{k};q\left|_{(\mu+\lambda,k),B^*}^{A^*,(\lambda,k)}\right.\right] = \sum_{R=0}^{\mu}\frac{(-1)^R q^{R(R+1)/2+\lambda R}\left(q^{-\mu};q\right)_R\left(q^{\lambda};q\right)_{\mu-R}}{(q;q)_R}$$

$$\overline{H}_{p+1,q+1}^{m+1,n}\left[\rho\left(zq^{\mu}\right)^{k};q\left|_{(R,k),B^*}^{A^*,(0,k)}\right.\right],$$

In chapter 4, Section 1 we established two new useful theorems for the generalized differential operator $D_{k,\alpha,x}^{m}$. As an application of our main results, we obtain two multiplication formulae for $\overline{H}$-function.

Theorem 1.

$$D^{m}_{l,\lambda-\mu,t}\left\{t^{\lambda-1}S_N^M[wt^{\rho}]f(xt)\right\}=\sum_{k=0}^{[N/M]}\frac{(-N)_{Mk}}{k!}A_{n,k}w^k\sum_{n=0}^{\infty}\frac{(-x)^n}{n!}$$

$$\prod_{p=0}^{m-1}\frac{\Gamma(\lambda+\rho k+pl)}{\Gamma(\mu+\rho k+pl)}t^{\lambda+\rho k+ml-1}{}_{m+1}F_m\left[{}^{-n,\lambda+\rho k,\lambda+\rho k+l,\ldots,\lambda+\rho k+(m-1)l}_{\mu+\rho k,\mu+\rho k+l,\ldots,\mu+\rho k+(m-1)l}\right]D_x^n\{f(x)\}$$

Theorem 2.

$$D^{m}_{l,\lambda-\mu,t}\left\{t^{\lambda}S_N^M[wt^{\rho}]f(xt)\right\}=\sum_{k=0}^{[N/M]}\frac{(-N)_{Mk}}{k!}A_{n,k}w^k$$

$$\sum_{n=0}^{\infty}\frac{(-x)^n}{n!}\prod_{p=0}^{m-1}\frac{\Gamma(\lambda+\rho k+pl)\Gamma(1-\mu-\rho k-pl)_n}{\Gamma(\mu+\rho k+pl)\Gamma(1-\lambda-\rho k-pl)_n}$$

$$t^{\lambda+\rho k+ml-1}{}_{m+1}F_m\left[{}^{-n,\lambda+\rho k-n,\lambda+\rho k-n+l,\ldots,\lambda+\rho k-n+(m-1)l}_{\mu+\rho k-n,\mu+\rho k-n+l,\ldots,\mu+\rho k-n+(m-1)l}\right]D_x^n\{f(x)\}$$

In section 2 We establish four transformations formulae of double infinite series involving the I -function. These formulas are then used to obtain double summation formulas for the said function. Our results are quite general in character and a number of summation formulas can be deduced as particular cases

First formula

$$\sum_{m,n=0}^{\infty}x^m y^n I^{m,n+2}_{p_i+2,q_i+1:r}\left[z\left|{}^{(1-a-m,\rho),(1-b-n,\sigma),A^*}_{B^*,(1-a-b-m-n,\sigma+\rho)}\right.\right]=(x+y-xy)^{-1}$$

$$\left\{x^{s+1}\sum_{s=0}^{\infty}x^{s+1}I^{m,n+2}_{p_i+2,q_i+1:r}\left[z\left|{}^{(1-a-s,\rho),(1-b,\sigma),A^*}_{B^*,(1-a-b-s,\sigma+\rho)}\right.\right]+\sum_{t=0}^{\infty}y^{t+1}I^{m,n+2}_{p_i+2,q_i+1:r}\left[z\left|{}^{(1-a,\rho),(1-b-t,\sigma),A^*}_{B^*,(1-a-b-t,\sigma+\rho)}\right.\right]\right\}$$

Second formula

$$\sum_{m,n=0}^{\infty}\frac{x^m y^n}{m!n!} I^{m,n}_{p_i,q_i:r}\left[z\left|\begin{matrix}(1-a-m-n,u),(1-b-m,v),A^*\\ B^*,(1-c-m,\omega)\end{matrix}\right.\right]$$

$$=\sum_{k=0}^{\infty}\frac{1}{k!}(1-y)^{-a}\left(\frac{x}{1-y}\right)^k I^{m,n+2}_{p_i+2,q_i+1:r}\left[z(1-y)^{-u}\left|\begin{matrix}(1-a-k,u),(1-b-k,v),A^*\\ B^*,(1-c-k,\omega)\end{matrix}\right.\right]$$

Third formula

$$\sum_{m,n=0}^{\infty}\frac{x^m y^n}{m!n!} I^{m,n+3}_{p_i+3,q_i+2:r}\left[z\left|\begin{matrix}(1-a-m-n,u),(1-b-m,v),(1-b'-n,\omega)A^*\\ B^*,(1-a-m,u),(1-a-n,u)\end{matrix}\right.\right]$$

$$=\sum_{k=0}^{\infty}\frac{1}{k!}(1-x)^{-b}(1-y)^{b'}\left(\frac{xy}{(1-x)(1-y)}\right)^k I^{m,n+2}_{p_i+2,q_i+1:r}\left[z(1-x)^{-v}(1-y)^{-u}\left|\begin{matrix}(1-b-k,v),(1-b'-k,\omega),A^*\\ B^*,(1-a-k,u)\end{matrix}\right.\right]$$

Fourth formula

$$\sum_{m,n=0}^{\infty}\frac{x^m y^n}{m!n!} I^{m,n+3}_{p_i+3,q_i+1:r}\left[z\left|\begin{matrix}(1-a-m-n,u),(1-b-m,v),(1-b'-n,\omega)A^*\\ B^*,(1-b-b'-m-n,\omega+v)\end{matrix}\right.\right]$$

$$=\sum_{k=0}^{\infty}(1-y)^{-a}\frac{1}{K!}\left(\frac{x-y}{1-y}\right)^k I^{m,n+3}_{p_i+3,q_i+1:r}\left[z(1-y)^{-u}\left|\begin{matrix}(1-a-k,u),(1-b-k,v),(1-b',\omega),A^*\\ B^*,(1-b-b'-k,\omega+v)\end{matrix}\right.\right]$$

In chapter 5, Section 1 we establish four interesting theorems exhibiting interconnections between images and originals of related functions in the Laplace transform. We also derive six corollaries of the theorems. Further, we obtain five new and general integrals by the application of the theorems

The I-function occurring in this section is defined and represented as follows:

$$I[z]=I^{m,n}_{p_i,q_i:r}[z]=I^{m,n}_{p_i,q_i:r}\left[z\left|\begin{matrix}(a_j,\alpha_j)_{1,n},(a_{ji},\alpha_{ji})_{n+1,p_i}\\(b_j,\beta_j)_{1,m},(b_{ji},\beta_{ji})_{m+1,q_i}\end{matrix}\right.\right]=\frac{1}{2\pi\omega}\int_L \phi(\xi)z^{\xi}\,d\xi$$

In section 2 we derive the solution of a convolution integral equation whose kernel is a generalized hypergeometric function p F Q[.] and the H-function of two variables. Some interesting special cases of main result have also been discussed.

The solution of the following convolution integral equation has been given:

$$\int_0^x (x-t)^{\sigma-1}\,{}_PF_Q\left[(g_P);(h_Q);a(x-t)^{\eta}\right]H^{o,n_1:m_2,n_2:m_3,n_3}_{p_1,q_1:p_2,q_2:p_3,q_3}\left[\begin{matrix}(x-t)\\(x-t)\end{matrix}\left|\begin{matrix}(a_j;\alpha_j,A_j)_{1,p_1}:(c_j,\gamma_j)_{1,p_2}:(e_j,E_j)_{1,p_3}\\(b_j;\beta_j,B_j)_{1,q_1}:(d_j,\delta_j)_{1,q_2}:(f_j,F_j)_{1,q_3}\end{matrix}\right.\right]$$
$$f(t)dt=g(x)$$

In chapter 6, Section 1 We derive three compositions of the fractional integral operators associated with a product of I-function and a general class of polynomials due to Srivastava.

following integral equations will be derived.

$$Y^h_\alpha\{f(x)\}=x^{-h-\alpha}A_\alpha\{x^h f(x)\}=\frac{x^{-h-\alpha}}{\Gamma(\alpha)}\int_0^x (x-s)^{\alpha-1}I^{m,n}_{p_i,q_i:r}\left[z\left(1-\frac{s}{x}\right)^c\right]$$
$$S^a_b\left[y\left(1-\frac{s}{x}\right)^{\sigma}\right]s^h f(s)\,ds$$

In Section 2 We have evaluated an integral involving an exponential function, Sine function, generalized hypergeometric series and I-function, and we have employed it to evaluate a double integral and establish Fourier series for the product of generalized hypergeometric functions. We have also derived a double Fourier exponential series for the I-function. Our results are unified in nature and act as a key formula from which we can derive many results as their particular cases

The following formula are required in the proofs:

$$\int_0^{\pi}(\sin x)^{\omega-1}e^{imx}\,{}_PF_Q[{}^{\alpha_P}_{\beta_Q};c(\sin x)^{2h}]dx=\frac{\pi e^{im\pi/2}}{2^{\omega-1}}\sum_{r=0}^{\infty}\frac{(\alpha_P)_r c^r\Gamma(\omega+2hr)}{(\beta_Q)_r r!2^{2hr}\Gamma(\frac{\omega+2hr\pm m+1}{2})}$$

In chapter 7 We present an analytic proof of the integrals for astrophysical thermonuclear functions which are derived on the basis of Boltzmann-Gibbs statistical mechanics. Among the four different cases of astrophysical thermonuclear functions, those with a depleted high-energy tail and a cut-off at high energies find a natural interpretation in q-statistics.

Mellin-Barnes integral representation for the exponential function, namely [Mathai and Saxena]

$$\exp[-x]=\frac{1}{2\pi\omega}\int_L\Gamma(s)x^{-s}ds,\ |x|<\infty$$

CHAPTER 1

A BRIEF SURVEY OF THE WORK DONE ON FRACTIONAL CALCULUS AND GENERALIZED INTEGRAL TRANSFORMATION

In this chapter we present a brief survey of the research work done on generalized hyper geometric function. Definitions of nearly all the generalized hyper geometric function and fractional integration operators are given. Some of which will be employed in presenting the results of the subsequent chapters. A brief survey of the researches carried out in the field of fractional integration operators is also presented.

1.1. INTRODUCTION

Special functions theory is well known in mathematics for its apparent multiplicity of tricks and devices needed to obtain information about the subject matter. A study of the transformation theory and properties of the generalized hyper geometric function enable one to derive the properties of a large class of Special functions in a unified manner.

There are three important aspects of the study of the generalized hyper geometric function which have attracted the attention of a majority of researches in this branch of analysis.

The first aspect deals with the investigations leading to certain generalizations of the question of summing up , in terms of product of gamma function only, particular types of generalized hyper geometric function and their extensions. The second aspect comprises of the study of the interrelations that might exist between such function. The third aspect deals with the generalizations of these functions which have recently led to many remarkable applications in physics, Statistics, Number theory and Combinatory Analysis.

It is well-known that in the study of the second order linear differential equation with three regular singular points, there arises the function

$$ {}_2F_1(a,b;c;z) = \sum_0^\infty \frac{(a)_n (b)_n}{(1)_n (c)_n} z^n \qquad (1.1.1) $$

Where c is neither zero nor a negative integer, where in (1.1.1.), the pochhammer's symbol employed is defined as

$(\alpha)_n = \alpha(\alpha+1)(\alpha+2).......(\alpha+n-1), n \geq 1$

$(\alpha)_0 = 1$

In the year 1908, Branes defined the hyper geometric function in terms of a Mellin-type integral deviating from the conventional method of defining a special function in terms of an infinite series, in the integral form:

$$ {}_2F_1(a,b,c;-z) = \frac{\Gamma(c)}{\Gamma(a)\Gamma(b)} \frac{1}{2\pi i} \int_{-i\infty}^{+i\infty} \frac{\Gamma(-s)\Gamma(a+s)\Gamma(b+s)}{\Gamma(c+s)} z^s ds \qquad (1.1.2) $$

Where $i = \sqrt{-1}$.

Where the pole of $\Gamma(-s)$, at the point s=0,1,2,3,..............., are separated from those of $\Gamma(a+s)$, at the points $s = -a - v_1 (v_1 = 0,1,2,......)$ and $\Gamma(b+s)$ at the points

$s = -b - v_2 (v_2 = 0,1,2,......)$ And

$|\arg(-z)| < \pi$

One of the importance of this definition lies in the fact that the Millen transform of ${}_2F_1(.)$ is the coefficient of z^{-s} in the integrand of (1.1.2.), that is

$$ \int_0^\infty z^{s-1} {}_2F_1(a,b,c;-z) dz = \frac{\Gamma(c)\Gamma(s)\Gamma(a-s)\Gamma(b-s)}{\Gamma(a)\Gamma(b)\Gamma(c-s)}, \qquad (1.1.3) $$

Where

$$ \mathrm{Re}(s) > 0, \mathrm{Re}(a-s) > 0, \mathrm{Re}(b-s) > 0. $$

Swaroop (1964) introduced and studied the hypergeometric function transform; that kernel is the Gauss's hyper geometric function. Saxena,(1967b) and Kalla and Saxena (1969) used the hypergeometric function in defining certain fractional integration operators. Mathai and saxena (1966) introduced the probability function associated with a ${}_2F_1$(a,b;c; .).

Generalized hypergeometric function is defined in the form:

$$ {}_pF_q\left[\begin{matrix}\alpha_1,\ldots,\alpha_p;\\ \beta_1,\ldots,\beta_q;\end{matrix} z\right] = F\left[\begin{matrix}(\alpha_p)\\ (\beta_q)\end{matrix}; z\right] = \sum_{n=0}^{\infty} \frac{\prod_{j=1}^{p}(\alpha_j)_n}{\prod_{j=1}^{q}(\beta_j)_n} \frac{z^n}{n!}; \qquad (1.1.4) $$

In which no denominator parameter is allowed to take zero or a negative integer value. If any of the parameter α_p is zero or a negative integer, the series terminates. The function defined by (1.1.4) will be denoted briefly by ${}_pF_q(z)$.

The Mellin-Branes integral for ${}_pF_q(z)$ is

$$ {}_pF_q\left[\begin{matrix}\alpha_1,\ldots,\alpha_p;\\ \beta_1,\ldots,\beta_q;\end{matrix} z\right] = $$

$$ \frac{\prod_{j=1}^{p}\Gamma(\beta_j)}{\prod_{j=1}^{q}\Gamma(\alpha_j)} \frac{1}{2\pi i} \int_{-i\infty}^{+i\infty} \frac{\prod_{j=1}^{p}\Gamma(\alpha_j+s)\Gamma(-s)}{\prod_{j=1}^{q}\Gamma(\beta_j+s)} (-z)^s\, ds \qquad (1.1.5) $$

Where $\mathrm{i} = \sqrt{-1}$ and for convergence, $\mathrm{p} \le \mathrm{q}$ or

$(\mathrm{p} = \mathrm{q} + 1 \text{ and } |\mathrm{z}| < 1)$, $|\arg(-\mathrm{z})| < \pi$

When $\mathrm{p} > \mathrm{q}+1$ the series in (1.1.4) diverges. The path of integration is indented, if necessary , in such a manner that the poles of $\Gamma(-s)$ at the points $s = 0, 1, 2, \ldots;$ are separated from those of $\Gamma(\alpha_j + s)$, at the points $\alpha_j = -s - v_j \;\; ; \;\; v_j = 0, 1, 2, \ldots ; \mathrm{j} = 1, \ldots, p$. An empty product is interpreted as unity.

It is a matter of common knowledge that the Gaussian hypergeometric function ${}_2F_1(\mathrm{a, b; c; 1})$ can be summed up as $\frac{\Gamma(c)\Gamma(c-a-b)}{\Gamma(c-a)\Gamma(c-b)}$, when $\mathrm{Re}(\mathrm{c}-\mathrm{a}-\mathrm{b}) > 0$ and that this formula is one of the simplest case of an assumable hypergeometric series which has found much use in the simplification of many problems.

In order to give a meaning to the symbol ${}_pF_q(.)$ When $\mathrm{p} > \mathrm{q}+1$, MacRobert (1937-1941) defined and studied his E-function in the form

$$E(p;a_r;q;\rho_B:x)=\frac{\Gamma(a_q+1)}{\Gamma(\rho_1-a_1)\Gamma(\rho_2-a_2)...\Gamma(\rho_q-a_q)}$$

$$.\prod_{u=1}^{q}\int_0^\infty \lambda_u^{\rho_u-a_u-1}(1+\lambda_u)^{-\rho_u}\,d\lambda_u$$

$$.\prod_{v=2}^{p-q-1}\int_0^\infty e^{-\lambda_{q+v}}\lambda_{q+v}^{a_{q+v}-1}\,d\lambda_{q+v}\int_0^\infty e^{-\lambda_p}$$

$$.\lambda_p^{a_p-1}\left[1+\frac{\lambda_{q+2}\lambda_{q+3}...\lambda_p}{(1+\lambda_1)(1+\lambda_2)...(1+\lambda_q)x}\right]^{-a_{q+1}}d\lambda_p \qquad (1.1.6)$$

A detailed account of this function can be found in Erdelyi et al. (1953).

Meijer (1946) introduced a generalization of the E- function in the form:

$$G_{p,q}^{m,n}\left[z\,|\,\begin{matrix}(a_p)\\(b_q)\end{matrix}\right]=G_{p,q}^{m,n}\left[z\,|\,\begin{matrix}a_1,...,a_p\\b_1,...,b_q\end{matrix}\right]$$

$$=\frac{1}{2\pi i}\int_L\frac{\prod_{j=1}^{m}\Gamma(b_j-s)\prod_{j=1}^{n}\Gamma(1-a_j+s)}{\prod_{j=m+1}^{q}\Gamma(1-b_j+s)\prod_{j=n+1}^{p}\Gamma(a_j-s)}z^s ds \qquad (1.1.7)$$

Where L is a suitable contour separating the poles of $\Gamma(b_j-s)$ for $j=1,....,m$

From those of $\Gamma(1-a_j+s)$ for $j=1,....,n$. The poles of theintegrand are assumed to be simple.

In an attempt to discover, the solution of certain integral equations Saxena, V. P. (1982) introduced the I - function in the following form:

$$I[z]=I_{p_i,q_i:r}^{m,n}[z]$$

$$=I_{p_i,qi;r}^{m,n}\left[z\,|\,\begin{matrix}\{(a_j,\alpha_j)_{1,n}\},\{(a_{ji},\alpha_{ji})_{n+1,p_i}\}\\\{(b_j,\beta_j)_{1,m}\},\{(b_{ji},\beta_{ji})_{m+1,q_i}\}\end{matrix}\right]$$

$$=\frac{1}{2\pi\omega}\int_L t(s)z^s ds \qquad (1.1.8)$$

Where $\omega=\sqrt{-1}$;

$$t(s)=\frac{\prod_{j=1}^{m}\Gamma(b_j-\beta_j s)\prod_{j=1}^{n}\Gamma(1-a_j+\alpha_j s)}{\sum_{i=1}^{r}\left\{\prod_{j=m+1}^{q_i}\Gamma(1-b_{ji}+\beta_{ji}s)\prod_{j=n+1}^{p_i}\Gamma(a_{ji}-\alpha_{ji}s\right\}} \tag{1.1.9}$$

$p_i, q_i (i = 1, \ldots, r)$,m,n are integers satisfying $0 \le n \le p_i$; $0 \le m \le q_i$ $(i = 1, \ldots, r)$, r is finite, $\alpha_j, \beta_j, \alpha_{ji}, \beta_{ji}$ are real and positive and a_j, b_j, a_{ji}, b_{ji} are complex numbers such that

$\alpha_j(b_h+\upsilon) \ne \beta_h(a_j-\upsilon-k)$, for υ, $k=1, 2, \ldots$; $h = 1, 2, \ldots, m$; $j = 1, 2, \ldots, n$.

L is a contour which runs from $\sigma - w^{\infty}$ to $\sigma + w^{\infty}$ (σ is real),

$$s = (a_j - 1 - \upsilon)/\alpha_j;\ j = 1, 2, \ldots, n;\ \upsilon = 0, 1, 2, \ldots$$

$$s = (b_j + \upsilon)/\beta_j;\ j = 1, 2, \ldots, m;\ \upsilon = 0, 1, 2, \ldots$$

Lie to the L.H.S. and R.H.S. of L, respectively.

I-function reduces to H-function, when $r = 1$.

The relation between I-and H-function is given below:

$$I^{m,n}_{p_i,q_i;1}\left[z\,\middle|\,\begin{matrix}\{(a_j,\alpha_j)_{1,n}\},\{(a_{ji},\alpha_{ji})_{n+1,p_i}\}\\ \{(b_j,\beta_j)_{1,m}\},\{(b_{ji},\beta_{ji})_{m+1,q_i}\}\end{matrix}\right]$$

$$=H^{m,n}_{p_1,q_1}\left[z\,\middle|\,\begin{matrix}(a_1,\alpha_1),(a_2,\alpha_2),\ldots,(a_{p_1},\alpha_{p_1})\\ (b_1,\beta_1),(b_2,\beta_2),\ldots,(b_{q_1},\beta_{q_1})\end{matrix}\right] \tag{1.1.10}$$

Viashya,Jain and Verma(1989) found certain identity, multiplication theorems differentiation formulae and some integrals involving the I-function (1.1.8).

Agarwal,(1965) extended the Meijer's G-function to G-function of two variables. The work of agarwal ,(1965) and Sharma, (1965) gave a fresh impetus to numerous workers to further generalize the G-symbol to G-function of n- variables due to Khadia(1970).The G-function of n- variables was further converted to the H-symbol of n- variables by Saxena (1974,77) further generalized the H- function of n- variables into the multivariable I- function .

In all these, aforesaid G -, H- and I- type function of one two and - n variables, the coefficients of the variable of integration in the gamma function products of the integrand (of the integrals defining the functions) were taken to be real positive. Considering these multipliers as complex number quite a few papers have appeared in the literature.

Gautam and Goyal (1981) defined the multivariable A- function , which is a generalization of multivariable H- function of Srivastava and Panda (1976) and belongs to the letter category of special functions.

The aforesaid generalised hypergeometric function have also been studied by Gupta and Rathie (1968), Khadia and Goyal (1975) , Love (1967) , Srivastava and Buschman (1972) , Bora and Saxena (1971) ,Barnes (1908), Bajpai and AL-Hawaj (1989) through various important results with generalized hypergeometric functions .

Pandey and Pandey (1985) have obtained power series expansion for the modified H- function of several variables, which by assigning suitable value of the parameters give rise to the power series expansions given by Lawricella and others.

In the present work, we have concentrated our study on multivariable I- function, multivariable A- function, and generalized H- function.

The definitions of the various generalized hypergeometric function of one and more variables are given below:

1.2. MULTIVARIABLE I-FUNCTION

The multivariable I- function introduced by Prasad (1986) will be defined and represented in the following manner:

$$I[z_1,\ldots,z_r] = I^{0,q_2:\ldots;0,n_r:(m',n');\ldots;(m^{(r)},n^{(r)})}_{p_2,q_2:\ldots;p_r,q_r:(p',q');\ldots;(p^{(r)},q^{(r)})}$$

$$\cdot\left[\begin{matrix} z_1 \\ \vdots \\ z_r \end{matrix}\left|\begin{matrix} (a_{2j};\alpha'_{2j};\alpha''_{2j})_{1,p_2}:\ldots:(a_{rj};\alpha'_{rj},\ldots,\alpha^{r}_{rj})_{1,p_r}:(a'_j,\alpha'_j)_{1,p'};\ldots;(a^{(r)}_j,\alpha^{(r)}_j)_{1,p^{(r)}} \\ (b_{2j};\beta'_{2j};\beta''_{2j})_{1,q_2}:\ldots:(b_{rj};\beta'_{rj},\ldots,\beta^{r}_{rj})_{1,q_r}:(b'_j,\beta'_j)_{1,q'};\ldots;(b^{(r)}_j,\beta^{(r)}_j)_{1,q^{(r)}} \end{matrix}\right.\right]$$

$$= \frac{1}{(2\pi\omega)^r} \int_{L_1} \dots \int_{L_r} \phi_1(s_1)\dots\phi_r(s_r)\psi(s_1,\dots,s_r)$$

$$.z_1^{s_1}\dots z_r^{s_r} ds_1\dots ds_r \tag{1.2.1}$$

Where $\omega = \sqrt{-1}$;

$$\phi_i(s_i) = \frac{\prod_{j=1}^{m^{(i)}} \Gamma(b_j^{(i)} - \beta_j^{(i)} s_{(i)}) \prod_{j=1}^{n^{(i)}} \Gamma(1 - a_j^{(i)} + \alpha_j^{(i)} s_{(i)})}{\prod_{j=m^{(i)}+1}^{q^{(i)}} \Gamma(1 - b_j^{(i)} + \beta_j^{(i)} s_{(i)}) \prod_{j=n^{(i)}+1}^{p^{(i)}} \Gamma(a_j^{(i)} - \alpha_j^{(i)} s_{(i)})};$$

$$\forall i \in \{1, \dots, r\} \tag{1.2.2}$$

$$\psi(s_1,\dots s_r) = \frac{\prod_{j=1}^{n_2} \Gamma(1 - a_{2j} + \sum_{i=1}^{2} \alpha_{2j}^{(i)} s_i) \prod_{j=1}^{n_3} \Gamma(1 - a_{3j} + \sum_{j=1}^{3} \alpha_{3j}^{(i)} s_i)}{\prod_{j=n_2+1}^{p_2} \Gamma(a_{2j} - \sum_{i=1}^{2} \alpha_{2j}^{(i)} s_i) \prod_{j=n_3+1}^{p_3} \Gamma(a_{3j} - \sum_{i=1}^{3} \alpha_{3j}^{(i)} s_i)}$$

$$\cdot \frac{\dots \prod_{j=1}^{n_r} \Gamma(1 - a_{rj} + \sum_{i=1}^{r} \alpha_{rj}^{(i)} s_i)}{\dots \prod_{j=n_r+1}^{p_r} \Gamma(a_{rj} - \sum_{i=1}^{r} \alpha_{rj}^{(i)} s_i) \prod_{j=1}^{q_2} \Gamma(1 - b_{2j} + \sum_{i=1}^{2} \beta_{2j}^{(i)} s_i)}$$

$$\frac{1}{\dots \prod_{j=1}^{q_r} \Gamma(1 - b_{rj} + \sum_{i=1}^{r} \beta_{rj}^{(i)} s_i)} \tag{1.2.3}$$

$\alpha_j^{(i)}, \beta_j^{(i)}, \alpha_{kj}^{(i)}, \beta_{kj}^{(i)}$ $(i = 1, \dots, r)(k = 2, \dots, r)$ are positive numbers, $a_j^{(i)}, b_j^{(i)}, (i = 1, \dots, r)$ a_{kj}, b_{kj} $(k = 2, \dots, r)$ are complex numbers and here $m^{(i)}, n^{(i)}, p^{(i)}, q^{(i)}$ $(i = 1, \dots, r)$ $, n_k, p_k, q_k$ $(k = 2, \dots, r)$ are non negative integers where $0 \le m^{(i)} \le q^{(i)}$, $0 \le n^{(i)} \le p^{(i)}$, $q_k \ge 0$, $0 \le n_k \le p_k$. Here (i) denotes the numbers of the contours.

L_i In the complex s_i-plane of the Mellin Barnes type which runs from $-w^{\infty} to + w^{\infty}$

With indentation, if necessary to ensure that all the poles of $\Gamma(b_j^{(i)} - \beta_j^{(i)} s_i)$ (j= 1, ... ,$m^{(i)}$) are separated from those of $\Gamma(1 - a_j^{(i)} + \alpha_j^{(i)} s_i)$ (j= 1, ... ,$n^{(i)}$), $\Gamma(1 - a_{2j} + \sum_{i=1}^{2} \alpha_{2j}^{(i)} s_i)$ (j= 1, ... ,n_2), ... ,

$\Gamma(1 - a_{rj} + \sum_{i=1}^{r} \alpha_{rj}^{(i)} s_i)$ (j= 1, ... ,n_r).

According to the asymptotic expansion of the gamma function, the counter integral (1.2.1) is absolutely convergent provided that

$$|\arg z_i| < \frac{1}{2}\pi U_i, U_i > 0 \;;\quad i=1,\ 2,\ \ldots,r \tag{1.2.4}$$

Where

$$\begin{aligned} U_i = \sum_{j=1}^{n^i} \alpha_j^{(i)} - \sum_{j=n^{(i)}+1}^{p^{(i)}} \alpha_j^{(i)} + \sum_{j=1}^{m^i} \beta_j^{(i)} - \sum_{j=m^{(i)}+1}^{q^{(i)}} \beta_j^{(i)} \\ +(\sum_{j=1}^{n_2} \alpha_{2j}^{(i)} - \sum_{j=n_2+1}^{p_2} \alpha_{2j}^{(i)}) + (\sum_{j=1}^{n_3} \alpha_{3j}^{(i)} - \sum_{j=n_3+1}^{p_3} \alpha_{3j}^{(i)}) \\ +\ldots+(\sum_{j=1}^{n_r} \alpha_{rj}^{(i)} - \sum_{j=n_r+1}^{p_r} \alpha_{rj}^{(i)}) \\ -(\sum_{j=1}^{q_2} \beta_{2j}^{(i)} + \sum_{j=1}^{q_3} \beta_{3j}^{(i)} + \ldots + \sum_{j=1}^{q_r} \beta_{rj}^{(i)}) \end{aligned} \tag{1.2.5}$$

The asymptotic expansion of the I-function has been discussed by Prasad (1986). His results run as follow:

$I[z_1,\ldots,z_r] = 0(|z_1|^{\alpha_1} \ldots |z_r|^{\alpha_r}), \max\{|z_1|,\ldots,|z_r|\} \to 0$

Where $\alpha_i = \min \mathrm{Re}(b_j^{(i)} / \beta_j^{(i)})$, $j=1,\ \ldots,m^{(i)}$; i=1, ... ,r (1.2.6)

And $I[z_1,\ldots,z_r] = 0(|z_1|^{\beta_1} \ldots |z_r|^{\beta_r}), \min\{|z_1|,\ldots,|z_r|\} \to \infty$

Where $\beta_i = \max \mathrm{Re}(\frac{a_j^{(i)} - 1}{\alpha_j^{(i)}})$; $j=1,\ \ldots,n^{(i)}$,i=1, ... ,r

$$n_2 = n_3 = \ldots = n_r = 0 \tag{1.2.7}$$

In the contracted notation, this function can be written in the following manner:

$$I[z_1,...,z_r] = I^{0,n_2:...:0,n_r:M^{(r)}}_{p_2,q_2:...:p_r,q_r:N^{(r)}} \left[\begin{matrix} z_1 \\ \vdots \\ z_r \end{matrix} \middle| \begin{matrix} P_r : P^{(r)} \\ Q_r : Q^{(r)} \end{matrix} \right]$$

$$= \frac{1}{(2\pi\omega)^r} \int_{L_1} .... \int_{L_r} \phi_1(s_1)...\phi_r(s_r)\psi(s_1,...,s_r)$$

$$.z_1^{s_1}...z_r^{s_r} ds_1...ds_r \qquad (1.2.8)$$

Where

$$M^{(r)} = (m',n');...;(m^{(r)},n^{(r)}); \qquad (1.2.9)$$

$$N^{(r)} = (p',q');...;(p^{(r)},q^{(r)}); \qquad (1.2.10)$$

$$P_r = (a_{2j};\alpha_{2j}',\alpha_{2j}'')_{1,p_2} :....: (a_{rj};\alpha_{rj}',....,\alpha_{rj}^{(r)})_{1,p_r}; \qquad (1.2.11)$$

$$Q_r = (b_{2j};\beta_{2j}',\beta_{2j}'')_{1,q_2} :....: (b_{rj};\beta_{rj}',....,\beta_{rj}^{(r)})_{1,q_r}; \qquad (1.2.12)$$

$$P^{(r)} = (a_j',\alpha_j')_{1,p'};...;(a_j^{(r)},\alpha_j^{(r)})_{1,p^{(r)}}; \qquad (1.2.13)$$

$$Q^{(r)} = (b_j',\beta_j')_{1,q'};...;(b_j^{(r)},\beta_j^{(r)})_{1,q^{(r)}}; \qquad (1.2.14)$$

And the conditions and notations are similar to those given explicitly with (1.2.1).

1.3. MULTIVARIABLE H-FUNCTION

When $n_2 = n_3 = ... = n_{r-1} = 0 = p_2 = p_3 = ... = p_{r-1}$ and $q_2 = q_3 = ... = q_{r-1} = 0$; multivariable I- function reduces to the H-function of several variables due to Srivastava and Panda (1976) defined in the following manner, which it self is a generalization of the H- function of several variables due to Saxena, (1974).

$$H[z_1,...,z_r] = H^{0,n:m_1,n_1;...;m_r,n_r}_{p,q:p_1,q_1;...;p_r,q_r}$$

$$\cdot\left[\begin{array}{c} z_1 \\ \vdots \\ z_r \end{array}\middle| \begin{array}{l} (a_j; A_j^{'}, ..., A_j^{(r)})_{1,p} : (c_j^{'}, C_j^{'})_{1,p_1}; ...; (c_j^{(r)}, C_j^{(r)})_{1,p_r} \\ (b_j; B_j^{'}, ..., B_j^{(r)})_{1,q} : (d_j^{'}, D_j^{'})_{1,q_1}; ...; (d_j^{(r)}, D_j^{(r)})_{1,q_r} \end{array}\right]$$

$$= \frac{1}{(2\pi\omega)^r} \int_{L_1} .... \int_{L_r} \Phi(s_1, ..., s_r) \theta_1(s_1) ... \theta_r(s_r) z_1^{s_1} ... z_r^{s_r} \, .ds_1 ... ds_r \qquad (1.3.1)$$

Where $\omega = \sqrt{-1}$;

$$\Phi(s_1, ..., s_r) = \frac{\prod_{j=1}^{n} \Gamma(1 - a_j + \sum_{i=1}^{r} A_j^{(i)} s_i)}{\prod_{j=n+1}^{p} \Gamma(a_j - \sum_{i=1}^{r} A_j^{(i)} s_i) \prod_{j=1}^{q} \Gamma(1 - b_j + \sum_{i=1}^{r} B_j^{(i)} s_i)} \qquad (1.3.2)$$

$$\theta_i(s_i) = \frac{\prod_{j=1}^{m_i} \Gamma(d_j^{(i)} - D_j^{(i)} s_i) \prod_{j=1}^{n_i} \Gamma(1 - c_j^{(i)} + C_j^{(i)} s_i)}{\prod_{j=m_i+1}^{q_i} \Gamma(1 - d_j^{(i)} - D_j^{(i)} s_i) \prod_{j=n_i+1}^{p_i} \Gamma(c_j^{(i)} - C_j^{(i)} s_i)} \; ,$$

$$\forall \mathrm{i} \in \{1, \; .... \; , \mathrm{r}\} \qquad (1.3.3)$$

In (1.3.1) the superscript (i) stands for the number of prime, e.g., $b^{(1)} = b'$, $b^{(2)} = b''$ and so on; and an empty product is interpreted as unity. Further it is assumed that the parameters

$$\left\{\begin{array}{l} a_j, \mathrm{j} = 1, ..., \mathrm{p}; c_j^{(i)}, \mathrm{j} = 1, ..., p_i; \\ b_j, \mathrm{j} = 1, ..., q; d_j^{(i)}, \mathrm{j} = 1, ..., q_i; \end{array}\right\}, \qquad \forall \mathrm{i} \in \{1, \; .... \; , \mathrm{r}\}$$

Are complex numbers, and the associated coefficients

$$\left\{\begin{array}{l} A_j^{(i)}, \mathrm{j} = 1, ..., \mathrm{p}; C_j^{(i)}, \mathrm{j} = 1, ..., p_i; \\ B_j^{(i)}, \mathrm{j} = 1, ..., q; D_j^{(i)}, \mathrm{j} = 1, ..., q_i; \end{array}\right\}, \qquad \forall \mathrm{i} \in \{1, \; .... \; , \mathrm{r}\}$$

Are positive real number such that

$$\Omega_i = \sum_{j=1}^{p} A_j^{(i)} + \sum_{j=1}^{p_i} C_j^{(i)} - \sum_{j=1}^{q} B_j^{(i)} - \sum_{j=1}^{q_i} D_j^{(i)} \le 0, \qquad (1.3.4)$$

and

$$\Lambda_i = -\sum_{j=n+1}^{p} A_j^{(i)} + \sum_{j=1}^{n_i} C_j^{(i)} - \sum_{j=n+1}^{p_i} C_j^{(i)} - \sum_{j=1}^{q} B_j^{(i)}$$

$$+\sum_{j=1}^{m_i} D_j^{(i)} - \sum_{j=m_i+1}^{q_i} D_j^{(i)} > 0, \quad \forall i \in \{1, \ldots, r\} \quad (1.3.5)$$

Where the integrals n, p, m_i, n_i, p_i and q_i are constrained by the inequalities $0 \le n \le p, q \ge 0, 1 \le m_i \le q_i$ and $0 \le n_i \le p_i, \forall i \in \{1, \ldots, r\}$ and the equalities in (1.3.4) hold for suitably restricted values of the complex variables $z_1, \ldots, z_r$. The poles of the integrand in (1. 3.1) are assumed to be simple. The contour L_i in the complex s_i-plane is of the Mellin-Branes type which runs from $\omega^{-\infty}$ to $\omega^{+\infty}$ with indentations, if necessary, to ensure that all the poles of $\Gamma(d_j^{(i)} - D_j^{(i)} s_i)$,

$j = 1, \ldots, m_i$, are separated from those of $\Gamma(1 - \tau_j^{(i)} + C_j^{(i)} s_i)$, $j = 1, \ldots, n_i$,

and $\Gamma(1 - a_j + \sum_{i=1}^{r} A_j^{(i)} s_i)$, $j = 1, \ldots, n$; $\forall i \in \{1, \ldots, r\}$

The multivariable H- function (1.3.1) converges absolutely under the conditions (1.3.5) for

$$|\arg z_i| < \frac{1}{2} \Lambda_i \pi, \quad \forall i \in \{1, \ldots, r\}, \quad (1.3.6)$$

The asymptotic expansion of algebraic order for the multivariable H-function, which will need in the analysis, is given below:

$$H[z_1, \ldots, z_r] = \left\{ \begin{array}{c} 0(|z_1|^{A_1} \ldots |z_r|^{A_r}), \max\{|z_1|, \ldots, |z_r|\} \to 0 \\ 0(|z_1|^{B_1} \ldots |z_r|^{B_r}), n = 0, \min\{|z_1|, \ldots, |z_r|\} \to 0 \end{array} \right\} \quad (1.3.7)$$

For $i = 1, \ldots, r$, with

$$\left\{ \begin{array}{l} A_i = \min \operatorname{Re}(d_j^{(i)} / D_j^{(i)}), j = 1, \ldots, m_i \\ B_i = \max \operatorname{Re}[(c_j^{(i)} - 1) / C_j^{(i)}], j = 1, \ldots, n_i \end{array} \right\} \quad (1.3.8)$$

Provided that each of the inequalities in (1.3.4), (1.3.5) and (1.3.6) hold.

If $A_j' = \ldots = A_j^{(r)}, j = 1, \ldots, p; \quad B_j' = \ldots = B_j^{(r)}, j = 1, \ldots, q$ in (1.3.1),we get a special multivariable H-function studied by Saxena, (1974). On the other hand, if all of the capital letters are chosen to be one, the H-function of several variables defined by (1.3.1) reduces to the corresponding G-function of several variables studied by Khadia and Goyal (1970).

Several authors made a systematic study of this function due to its general character notably by Srivastava and Panda(1975, a, 1976,a, b, 78, 79), Panda (1977, a, b), Tondon (1980, a, b),Agrawal,(1980a),Goyal and Agrawal (1980, 81), Prasad and Singh (1977) , Siddiqui (1979), Agal and koul (1981), Buschman (1979), Dixit (1981), garg, (1980), Garg, (1981), Joshi and Arya (1981), Mathur, (1981), Srivastava, Koul and Raina (1981), Srivastava and Raina (1981), Srivastava and Singh (1981), Srivastava and Srivastava (1978), Srivastava,R. (1981),Munot and Mathur (1983), Saxena and Agarwal (1985) Chaurasia and Taygi (1991), Chaurasia (1988, 91), Gupta and Agrawal (1989) and others.

Generalized H-function as a symmetrical Fourier Kernel has been studied by Saxena and Modi (1975).

1.4. H-FUNCTION OF THREE VARIABLES

When $r = 3$, (1.3.1) reduces to the H-function of three variables defined and represented by means of triple Mellin-Barnes integral in the form:

$$H\begin{bmatrix} x \\ y \\ z \end{bmatrix} = H^{0,n:m_1,n_1;m_2,n_2;m_3,n_3}_{p,q:p_1,q_1;p_2,q_2;p_3,q_3}$$

$$\cdot\left[\begin{matrix} x \\ y \\ z \end{matrix}\middle| \begin{matrix} (a_j;A_j',A_j'',A_j''')_{1,p}:(c_j',C_j')_{1,p_1};(c_j'',C_j'')_{1,p_2};(c_j''',C_j''')_{1,p_3} \\ (b_j;B_j',B_j'',B_j''')_{1,q}:(d_j',D_j')_{1,q_1};(d_j'',D_j'')_{1,q_2};(d_j''',D_j''')_{1,q_3} \end{matrix}\right]$$

$$= \frac{1}{(2\pi\omega)^3} \int_{L_1}\int_{L_2}\int_{L_3} \Phi(s_1,s_2,s_3)\theta_1(s_1)\theta_2(s_2)\theta_3(s_3)$$

$$.x^{s_1} y^{s_2} z^{s_3} ds_1 ds_2 ds_3 \tag{1.4.1}$$

Where $\omega = \sqrt{-1}$;

$$\Phi(s_1, s_2, s_3) = \frac{\prod_{j=1}^{n} \Gamma(1 - a_j + \sum_{i=1}^{3} A_j^{(i)} s_i)}{\prod_{j=n+1}^{p} \Gamma(a_j - \sum_{i=1}^{3} A_j^{(i)} s_i) \prod_{j=1}^{q} \Gamma(1 - b_j + \sum_{i=1}^{3} B_j^{(i)} s_i)}; \tag{1.4.2}$$

$$\theta_i(s_i) = \frac{\prod_{j=1}^{m_i} \Gamma(d_j^{(i)} - D_j^{(i)} s_i) \prod_{j=1}^{n_i} \Gamma(1 - \tau_j^{(i)} + C_j^{(i)} s_i)}{\prod_{j=m_i+1}^{q_i} \Gamma(1 - d_j^{(i)} + D_j^{(i)} s_i) \prod_{j=n_i+1}^{p_i} \Gamma(\tau_j^{(i)} - C_j^{(i)} s_i)},$$

$$\forall i \in \{1,\ 2,\ 3\}. \tag{1.4.3}$$

The conditions of existence of the H-function of three variables can be obtained from (1.3.4), (1.3.5) and (1.3.6) on setting $r = 3$.

1.5. H-FUNCTION OF TWO VARIABLES

Mittal and Gupta (1972) defined the H-function of two variables. In the notation of Srivastava and Panda (1976), the H-function of two variables is defined and represented by means of double Mellin-Barnes integral in the form:

$$H\begin{bmatrix} x \\ y \end{bmatrix} = H^{0,n:m_1,n_1;m_2,n_2}_{p,q:p_1,q_1;p_2,q_2} \cdot \left[\begin{matrix} x \\ y \end{matrix} \middle| \begin{matrix} (a_j; A_j', A_j'')_{1,p} : (c_j', C_j')_{1,p_1}; (c_j'', C_j'')_{1,p_2} \\ (b_j; B_j', B_j'')_{1,q} : (d_j', D_j')_{1,q_1}; (d_j'', D_j'')_{1,q_2} \end{matrix} \right]$$

$$= -\frac{1}{4\pi^2} \int_{L_1} \int_{L_2} \Phi(s_1, s_2) \theta_1(s_1) \theta_2(s_2) x^{s_1} y^{s_2} ds_1 ds_2 \tag{1.5.1}$$

The functions $\phi(\xi, \eta)$ and $[\theta_1(\xi)$ and $\theta_2(\eta)]$ Can be obtained on setting $r = 2$ in (1.3.2) and (1.3.3) respectively.

We mention below some interesting and useful special cases of the H-function of two variables.

(i) If $A_j' = A_j''(j = 1, \ldots, p), B_j' = B_j''(j = 1, \ldots, q)$ in (1.5.1),We obtain the special H-function of two variables studied by a number of workers such as

Munot and Kalla(1971), Bora and Kalla(1970), Chaturvedi and Goyal(1972), Saxena, (1971,b),Pathak(1970),Shah(1973),Verna,(1971) and others.

(ii) If we assume all capital letters with their dashes as unity, we obtain a relationship of H-function of two variables and G-function of two variables.

$$H^{0,n:m_1,n_1;m_2,n_2}_{p,q:p_1,q_1;p_2,q_2}\left[\begin{matrix} x \\ y \end{matrix}\Big| \begin{matrix} (a_j;1,1)_{1,p} : (c_j^{'},1)_{1,p_1} ; (c_j^{"},)_{1,p_2} \\ (b_j;1,1)_{1,q} : (d_j^{'},1)_{1,q_1} ; (d_j^{"},1)_{1,q_2} \end{matrix}\right]$$

$$= G^{0,n:m_1,n_1;m_2,n_2}_{p,q:p_1,q_1;p_2,q_2}\left[\begin{matrix} x \\ y \end{matrix}\Big| \begin{matrix} (a_p) : (c_{p_1}^{'}) ; (c_{p_2}^{"}) \\ (b_q) : (d_{q_1}^{'}) ; (d_{q_2}^{"}) \end{matrix}\right] \tag{1.5.2}$$

The G-function of two variables appearing on the R.H.S. of (1.5.2) was introduced by Agarwal,(1965). In the notation of Srivastava and Joshi [(1969), p. 471], $G[x,y]$ is represented as follows:

$$G\begin{bmatrix} x \\ y \end{bmatrix} = G^{0,n:m_1,n_1;m_2,n_2}_{p,q:p_1,q_1;p_2,q_2}\left[\begin{matrix} x \\ y \end{matrix}\Big| \begin{matrix} (a_p) : (c_{p_1}^{'}) ; (c_{p_2}^{"}) \\ (b_q) : (d_{q_1}^{'}) ; (d_{q_2}^{"}) \end{matrix}\right]$$

$$= -\frac{1}{4\pi i}\int_{L_1}\int_{L_2}\Phi(\xi+\eta)\psi_1(\xi)\psi_2(\eta)x^{\xi}y^{\eta}d\xi d\eta \ , \tag{1.5.3}$$

Where an empty product is interpreted as unity,

$$\Phi(\rho) = \frac{\prod_{j=1}^{n}\Gamma(1-a_j+\rho)}{\prod_{j=n+1}^{p}\Gamma(a_j-\rho)\prod_{j=1}^{q}\Gamma(1-b_j+\rho)} \tag{1.5.4}$$

$$\psi_1(\xi) = \frac{\prod_{j=1}^{m_1}\Gamma(d_j^{'}-\xi)\prod_{j=1}^{n_1}\Gamma(1-\tau_j^{'}+\xi)}{\prod_{j=m_1+1}^{q_1}\Gamma(1-d_j^{'}+\xi)\prod_{j=n_1+1}^{p_1}\Gamma(\tau_j^{'}-\xi)} \tag{1.5.5}$$

And with $\psi_2(\eta)$ defined analogously to $\psi_1(\xi)$ in terms of the parameter sets $(\tau_{p_2}^{"})$ and $(d_{q_2}^{"})$. Here x and y are not equal to zero, p_i, q_i, n_i and m_i, p, q, n are non –negative integers such that $p \ge n \ge 0, q \ge 0, p_i \ge n_i \ge 0,\ q_i \ge m_i \ge 0\ ,(i=1,2)$

For the details of this function one can refer to the original papers by Agarwal, (1965).

1.6. GENERALIZED KAMPE DE FERIET FUNCTION

When

$n = p, z_i = -z_i, m_i = 1, n_i = p_i, q_i = q_i + 1, a_j = 1 - a_j, b_k = 1 - b_k$

$(j = 1, \ldots, p; k = 1, \ldots, q)$,

$c_g^{(i)} = 1 - c_g^{(i)}, d_h^{(i)} = 1 - d_h^{(i)} (g = 1, \ldots, p_i; h = 1, \ldots, q_i), \quad \forall i \in \{1, \ldots, r\}$

in multivariable H-function (1.3.1), an interesting relationship obtained as

$$H^{0,p:1,p_1;1,p_2;\ldots;1,p_r}_{p,q:p_1,q_1+1;p_2,q_2+1;\ldots;p_r,q_r+1}$$

$$\cdot\left[\begin{matrix}-z_1\\ \vdots\\ -z_r\end{matrix}\left|\begin{matrix}(1-a_j;A_j^{'},\ldots,A_j^{(r)})_{1,p}:(1-c_j^{'},C_j^{'})_{1,p_1};\ldots;(1-c_j^{(r)},C_j^{(r)})_{1,p_r}\\ (1-b_j;B_j^{'},\ldots,B_j^{(r)})_{1,q}:(1-d_j^{'},D_j^{'})_{1,q_1};\ldots;(1-d_j^{(r)},D_j^{(r)})_{1,q_r}\end{matrix}\right.\right]$$

$$= S^{p:p_1;p_2;\ldots;p_r}_{q:q_1;q_2;\ldots;p_r}$$

$$\cdot\left[\begin{matrix}(a_j;A_j^{'},\ldots,A_j^{(r)})_{1,p}:(c_j^{'},C_j^{'})_{1,p_1};\ldots;(c_j^{(r)},C_j^{(r)})_{1,p_r};\\ (b_j;B_j^{'},\ldots,B_j^{(r)})_{1,q}:(d_j^{'},D_j^{'})_{1,q_1};\ldots;(d_j^{(r)},D_j^{(r)})_{1,q_r};\end{matrix}\; z_1,\ldots,z_r\right] \tag{1.6.1}$$

$$= \; S[z_1 \ldots z_r]$$

Where $S[z_1 \ldots z_r]$ is the generalized Kampe de Feriet function of variables defined and represented as follows:

$$S[z_1,\ldots,z_r] = \sum_{s_1,\ldots,s_r=0}^{\infty} \Lambda(s_1,\ldots,s_r)\prod_{i=1}^{r}\{\theta_i(s_i)\frac{z_i^{s_i}}{(s_i)!}\} \tag{1.6.2}$$

Where, for the convenience,

$$\Lambda(s_1,\ldots,s_r) = \frac{\prod_{j=1}^{p}\Gamma(a_j + \sum_{i=1}^{r} A_j^{(i)} s_i)}{\prod_{j=1}^{q}\Gamma(b_j + \sum_{i=1}^{r} B_j^{(i)} s_i)} \tag{1.6.3}$$

And

$$\theta_i(s_i) = \frac{\prod_{j=1}^{p_i} \Gamma(\tau_j^{(i)} + C_j^{(i)} s_i)}{\prod_{j=1}^{q_i} \Gamma(d_j^{(i)} + D_j^{(i)} s_i)} \qquad \forall i \in \{1, \ldots, r\} \qquad (1.6.4)$$

The r –tuple series given by (1.6.2) converges absolutely, if

$$1 + \sum_{j=1}^{q} B_j^{(i)} + \sum_{j=1}^{q_i} D_j^{(i)} - \sum_{j=1}^{p} A_j^{(i)} - \sum_{j=1}^{p_i} C_j^{(i)} \geq 0\,;$$

$$\forall i \in \{1, \ldots, r\} \qquad (1.6.5)$$

Where each of the equalities holds when the variables are suitably constrained.

If we set $r = 2$, then (1.6.2), reduces to the generalized Kampe de Feriet function of two variables defined and studied by Srivastava and Daust (1969), represented as

$$S[x, y] = \sum_{s_1, s_2 = 0}^{\infty} \xi(s_1, s_2)\theta_1(s_1)\theta_2(s_2) \frac{z_1^{s_1} z_2^{s_2}}{s_1! s_2!} \qquad (1.6.6)$$

Where $\xi(s_1, s_2)$ and $\theta_i(s_i)$ $(i = 1, 2)$ can be easily found by (1.6.3) and (1.6.4) respectively on setting $r = 2$. The double series involved in (1.6.6) convergent if

$$1 + \sum_{j=1}^{q} B_j^{(i)} + \sum_{j=1}^{q_i} D_j^{(i)} - \sum_{j=1}^{p} A_j^{(i)} - \sum_{j=1}^{p_i} C_j^{(i)} \geq 0\,,$$

$$(i = 1, 2) \qquad (1.6.7)$$

Further, if we take all capital letters with dashes in (1.6.6) equal to unity, it reduces to the Kampe de Feriet function.

1.7. FOX'S H-FUNCTION

When $\mathrm{n} = \mathrm{p} = \mathrm{q} = 0$, (1.3.1); the multivariable H-function break up into the product of r Fox's H-function.

$$H^{0,0:m_1,n_1;\ldots;m_r,n_r}_{p,q:p_1,q_1;\ldots;p_r,q_r}\left[\begin{matrix} z_1 \\ \vdots \\ z_r \end{matrix} \Big| \begin{matrix} -:(c'_j,C'_j)_{1,p_1};\ldots;(c^{(r)}_j,C^{(r)}_j)_{1,p_r} \\ -:(d'_j,D'_j)_{1,q_1};\ldots;(d^{(r)}_j,D^{(r)}_j)_{1,q_r} \end{matrix}\right]$$

$$=\prod_{i=1}^{r}\left\{H^{m_i,n_i}_{p_i,q_i}\left[z_i \Big| \begin{matrix} (c^{(i)}_j,C^{(i)}_j)_{1,p_i} \\ (d^{(i)}_j,D^{(i)}_j)_{1,q_i} \end{matrix}\right]\right\} \quad (1.7.1)$$

Fox (1961) has introduced the H-function in the field of special function while investigating the most generalized Fourier Kernel in one variable. The H-function is defined in terms of Mellin-Branes type integral as

$$H^{m,n}_{p,q}\left[z\Big|\begin{matrix}(a_p,A_p)\\(b_q,B_q)\end{matrix}\right]=H^{m,n}_{p,q}\left[z\Big|\begin{matrix}(a_1,A_1),\ldots,(a_p,A_p)\\(b_1,B_1),\ldots,(b_q,B_q)\end{matrix}\right]$$

$$=\frac{1}{2\pi\omega}\int_L \chi(s)z^s ds \quad (1.7.2)$$

Where $\omega=\sqrt{-1}$;

$$\chi(s)=\frac{\prod_{j=1}^{m}\Gamma(b_j-\beta_j s)\prod_{j=1}^{n}\Gamma(1-a_j+\alpha_j s)}{\prod_{j=m+1}^{q}\Gamma(1-b_j+\beta_j s)\prod_{j=n+1}^{p}\Gamma(a_j-\alpha_j s)} \quad (1.7.3)$$

Where an empty product is interpreted as unity, $1 \le m \le q$; $0 \le n \le p$; A's and B's are all positive numbers, L is a contour of Barnes type such that the poles of $\Gamma(b_j-\beta_j s)$, for $j=1,\ldots,m$, are to its right, and those of $\Gamma(1-a_j+A_j s)$, $j=1,\ldots,n$ to the left of the contour L. The poles of the integrand are assumed to be simple. The integral converges if $|\arg z|<\frac{\pi}{2}D$, where

$$D=\sum_{j=1}^{n}A_j-\sum_{j=n+1}^{p}A_j+\sum_{j=1}^{m}B_j-\sum_{j=m+1}^{q}B_j>0 \quad (1.7.4)$$

And

$$\mu = \sum_{j=1}^{q} B_j - \sum_{j=1}^{p} A_j \geq 0 \qquad (1.7.5)$$

A detailed account of the analytic continuation and asymptotic expansion of the H-function has been given by Braaksma (1963).

G - and H-function have found a large number of applications in Mathematical physics and chemistry,biological, sociological and statistical sciences. In this connection, one can refer to the monographs by Mathai and Saxena (1973, 78).

Fox (1965) developed the formal solution of dual integral equations associated with a specialized H-function as its kernel by the application of fractional integration operators due to Erdelyi. His result has been generalized by Saxena, in a series of papers (1967, a), in which the formal solutions of dual integral equations associated with the H-function have been derived. Kumbhat and Saxena (1974) obtained formal solution of certain triple integral equations involving H-functions.

Gupta,K.C. (1965) evaluated some integrals involving Bessel, Whittaker and H-functions. Gupta and Jain (1966) also evaluated an integral of product of two H-functions generalizing Saxena's formula for the integral of product of two G-functions (1960). Goyal (1970) has evaluated some finite integrals involving the H-function. Anandani (1969, a, b) evaluated integrals associated with generalized associated Legendre function and the H-function.

Gupta, (1965), Jain, (1968) and Rathie (1979, 80) evaluated certain integrals involving H-function.

Expansion theorems for the H-function have been given by Lawrynowicz (1969), Skikinski (1970), Gupta and Jain (1966, 69), Gupta and Srivastava (1972) and others. Series representations of H-function are given by Jain, (1969). Srivastava and Daoust (1969) have derived some generalized Neumann's expansions associated with Kampe de Feriet function. Summation formulae and certain recurrence relations for the H-function have been proved by Jain, (1967) and others.

Bajpai (1971, 80), Sharma, (1965) and Shah (1969) have given certain series expansions of H-function in terms of orthogonal polynomails and the H-function.

An integral transformation associated with the H-function is defined and stuided by Gupta and Mittal (1970, 71) and Rattan singh (1968,70).

Mathai and Saxena (1966) used the H-function in the study of certain statistical distributions. A detailed account of the applications of H-function in statistical distributions is available from the monograph by Mathai and Saxena (1978).

Nair and Samar (1971) obtained the differential properties of the H-function. Saxena and Kumbhat (1974) defined certain operators of fractional integration associated with H-function. Buschman (1972) derived some relations of contiguity for the H-function.

1.8. THE MULTIVARIABLE A-FUNCTION

Gautam and Goyal (1981, 82) defined the multivariable A-function, which is a generalization of multivariable H-function of Srivastava and Panda (1976).

The definition of multivariable A-function runs as follows:

$$A[z_1,...,z_r] = A^{m,n:m_1,n_1;...;m_r,n_r}_{p,q:p_1,q_1;...;p_r,q_r}$$

$$\cdot\left[\begin{matrix} z_1 \\ \vdots \\ z_r \end{matrix}\,\middle|\, \begin{matrix} (a_j;A'_j,...,A^{(r)}_j)_{1,p};(c'_j,C'_j)_{1,p_1};...;(c^{(r)}_j,C^{(r)}_j)_{1,p_r} \\ (b_j;B'_j,...,B^{(r)}_j)_{1,q};(d'_j,D'_j)_{1,q_1};...;(d^{(r)}_j,D^{(r)}_j)_{1,q_r} \end{matrix}\right]$$

$$= \frac{1}{(2\pi\omega)^r}\int_{L_1}....\int_{L_r}\theta_1(s_1)....\theta_r(s_r)\Phi(s_1,....,s_r)z_1^{s_1}...z_r^{s_r}\,.ds_1...ds_r \tag{1.8.1}$$

Where $\omega = \sqrt{-1}$;

$$\theta_i(s_i)=\frac{\prod_{j=1}^{m_i}\Gamma(d_j^{(i)}-D_j^{(i)}s_i)\prod_{j=1}^{n_i}\Gamma(1-c_j^{(i)}+C_j^{(i)}s_i)}{\prod_{j=m_i+1}^{q_i}\Gamma(1-d_j^{(i)}+D_j^{(i)}s_i)\prod_{j=n_i+1}^{p_i}\Gamma(c_j^{(i)}-C_j^{(i)}s_i)}$$

$$\forall i\in\{1,\ .\ .\ .\ .\ ,\mathrm{r}\} \qquad (1.8.2)$$

$$\Phi(s_1,\dots,s_r)=\frac{\prod_{j=1}^{n}\Gamma(1-a_j+\sum_{i=1}^{r}A_j^{(i)}s_i)\prod_{j=1}^{m}\Gamma(b_j-\sum_{i=1}^{r}B_j^{(i)}s_i)}{\prod_{j=n+1}^{p}\Gamma(a_j-\sum_{i=1}^{r}A_j^{(i)}s_i)\prod_{j=m+1}^{q}\Gamma(1-b_j+\sum_{i=1}^{r}B_j^{(i)}s_i)} \qquad (1.8.3)$$

Here m , n , p , q , m_i , n_i , p_i , and $\mathrm{q}_i\left(\mathrm{i}=1,\ .\ .\ .\ .\ ,\mathrm{r}\right)$ are non-negative integers and all $a_j^{'}s, b_j^{'}s, d_j^{(i)}{'}s, c_j^{(i)}{'}s, A_j^{(i)}{'}s$ and $B_j^{(i)}{'}s$ are complex numbers.

The multiple integral defining the A-function of r-variables converges absolutely if

$$\left|\arg(\Omega_i)z_k\right|<\frac{\pi}{2}\eta_i,\xi_i^*=0,\eta_i>0 \qquad (1.8.4)$$

$$\Omega_i=\prod_{j=1}^{p}\{A_j^{(i)}\}^{A_j^{(i)}}\prod_{j=1}^{q}\{B_j^{(i)}\}^{-B_j^{(i)}}\prod_{j=1}^{q_i}\{D_j^{(i)}\}^{D_j^{(i)}}.\prod_{j=1}^{p_i}\{C_j^{(i)}\}^{-C_j^{(i)}}$$

$$,\ \forall i\in\{1,\ .\ .\ .\ .\ ,\mathrm{r}\}; \qquad (1.8.5)$$

$$\xi^*{}_i=I_m\left[\sum_{j=1}^{p}A_j^{(i)}-\sum_{j=1}^{q}B_j^{(i)}+\sum_{j=1}^{q_i}D_j^{(i)}-\sum_{j=1}^{q_i}C_j^{(i)}\right],\ \forall i\in\{1,\ .\ .\ .\ .\ ,\mathrm{r}\} \qquad (1.8.5)$$

$$\eta_i=\mathrm{Re}\left[\sum_{j=1}^{n}A_j^{(i)}-\sum_{j=n+1}^{p}A_j^{(i)}+\sum_{j=1}^{m}B_j^{(i)}-\sum_{j=m+1}^{q}B_j^{(i)}+\sum_{j=1}^{m_i}D_j^{(i)}-\sum_{j=m_i+1}^{q_i}D_j^{(i)}+\sum_{j=1}^{n_i}C_j^{(i)}-\sum_{j=n_i+1}^{p_i}C_j^{(i)}\right]$$

$$\forall i\in\{1,\ .\ .\ .\ .\ ,\mathrm{r}\}; \qquad (1.8.7)$$

If we take $A_j^{'}s, B_j^{'}s, C_j^{'}s$ and $D_j^{'}s$ as real and positive and $\mathrm{m}\ =\ 0$, the A-function reduces to multivariable H-function of Srivastava and Panda (1976), i.e. (1. 3. 1)

We are using the multivariable A-function defined by (1. 8. 1) in the following concise form throughout the text.

$$A[z_1,...,z_r] = A^{m,n:M_r}_{p,q:N_r}\left[\begin{matrix} z_1 \\ \vdots \\ z_r \end{matrix} \middle| \begin{matrix} P:P_r^{(r)} \\ Q:Q_r^{(r)} \end{matrix}\right]$$

$$= \frac{1}{(2\pi\omega)^r}\int_{L_1}....\int_{L_r}\theta_1(s_1)....\theta_r(s_r)\Phi(s_1,....,s_r)z_1^{s_1}...z_r^{s_r}.ds_1...ds_r$$

(1.8. 8)

Where $\omega = \sqrt{-1}$;

$M_r = m_1, n_1;...;m_r, n_r;$

$N_r = p_1, q_1;...;p_r, q_r;$

$P = (a_j; A_j^{'},..., A_j^{r})_{1,p};$

$Q = (b_j; B_j^{'},..., B_j^{r})_{1,q};$

$P_r^{(r)} = (c_j^{'}, C_j^{'})_{1,p_1};...;(c_j^{(r)}, C_j^{(r)})_{1,p_r};$

And the definition of the functions $\theta_i(s_i)$ i =1, ... ,r ; $\Phi(s_1,....,s_r)$ and the condition of existence of the multivariable A-function are the same as mentioned by Gautam and Goyal (1981).

1.9. THE GENERAL TRIPLES HPERGEOMETRIC SERIES $F^{(3)}[x,y,z]$

Following srivastava,[(1967), p.428], a general triple hypergeometric series $F^{(3)}[x,y,z]$ is defined as:

$$F^{(3)}[x,y,z] = F^{(3)}\left[\begin{matrix}(a)::(b);(b^{'});(b^{"}):(c);(c^{'});(c^{"}); \\ (e)::(g);(g^{'});(g^{"}):(h);(h^{'});(h^{"});\end{matrix} x,y,z\right]$$

$$= \sum_{m,n,p=0}^{\infty} \Lambda(m,n,p)\frac{x^m}{m!}\frac{y^n}{n!}\frac{z^p}{p!} \qquad (1.9.1)$$

Where for convenience,

$$\Lambda(m,n,p) = \frac{\prod_{j=1}^{A}(a_j)_{m+n+p}\prod_{j=1}^{B}(b_j)_{m+n}\prod_{j=1}^{B^{'}}(b^{'}_j)_{n+p}\prod_{j=1}^{B^{"}}(b^{"}_j)_{p+m}}{\prod_{j=1}^{E}(e_j)_{m+n+p}\prod_{j=1}^{G}(g_j)_{m+n}\prod_{j=1}^{G^{'}}(g^{'}_j)_{n+p}\prod_{j=1}^{G^{"}}(g^{"}_j)_{p+m}}$$

$$\cdot\frac{\prod_{j=1}^{C}(c_j)_m\prod_{j=1}^{C'}(c'_j)_n\prod_{j=1}^{C''}(c''_j)_p}{\prod_{j=1}^{H}(h_j)_m\prod_{j=1}^{H'}(h'_j)_n\prod_{j=1}^{H''}(h''_j)_p} \tag{1.9.2}$$

The triple hypergeometric series $F^{(3)}[x,y,z]$ defined by (1.9.1), converges absolutely, when

$$1+E+G+G''+H-A-B-B''-C\geq 0;$$

$$1+E+G+G'+H'-A-B-B'-C'\geq 0;$$

$$1+E+G'+G''+H''-A-B'-B''-C''\geq 0.$$

1.10. THE $\overline{H}$-FUNCTION

The $\overline{H}$-function occurring in the paper will be defined and represented as follows:

$$\overline{H}_{P,Q}^{M,N}\left[z\right]=\overline{H}_{P,Q}^{M,N}\left[z\left|\begin{matrix}(a_j;\alpha_j;A_j)_{1,N},(a_j;\alpha_j)_{N+1,P}\\(b_j,\beta_j)_{1,M},(b_j,\beta_j;B_j)_{M+1,Q}\end{matrix}\right.\right]=\frac{1}{2\pi i}\int_{-i\infty}^{i\infty}\overline{\phi}(\xi)z^{\xi}d\xi \tag{1.10.1}$$

Where $$\overline{\phi}(\xi)=\frac{\prod_{j=1}^{M}\Gamma(b_j-\beta_j\xi)\prod_{j=1}^{N}\left\{\Gamma(1-a_j+\alpha_j\xi)\right\}^{A_j}}{\prod_{j=M+1}^{Q}\left\{\Gamma(1-b_j+\beta_j\xi)\right\}^{B_j}\prod_{j=N+1}^{P}\Gamma(a_j-\alpha_j\xi)} \tag{1.10.2}$$

Which contains fractional powers of the gamma functions.Here, and Through out the paper $a_j(j=1,...,p)$ and $b_j(j=1,...,Q)$ are complex parameters $\alpha_j\geq 0(j=1,...,P),\beta_j\geq 0(j=1,...,Q)$ (not all zero simultaneously) and exponents $A_j(j=1,...,N)$ and $B_j(j=N+1,...,Q)$ can take on non integer values.

The following sufficient condition for the absolute convergence of the defining integral for the $\overline{H}$-function given by equation (1.1) have been given by Buschman and Srivastava (1990).

$$\Omega\equiv\sum_{j=1}^{M}\left|\beta_j\right|+\sum_{j=1}^{N}\left|A_j\alpha_j\right|-\sum_{j=M+1}^{Q}\left|\beta_jB_j\right|-\sum_{j=N+1}^{P}\left|\alpha_j\right|>0 \tag{1.10.3}$$

And $|\arg(z)| < \frac{1}{2}\pi\Omega$ (1.10.4)

The behavior of the $\overline{H}$-function for small values of $|z|$ follows easily from a result recently given by Rathie.

We have

$$\overline{H}_{P,Q}^{M,N}[z] = 0\left(|z|^{\gamma}\right), \gamma = \min_{1\le j\le N}\left[\operatorname{Re}\left(b_j/\beta_j\right)\right], |z| \to 0 \tag{1.10.5}$$

If we take $A_j = 1(j = 1,\ldots,N)$, $B_j = 1(j = M+1,\ldots,Q)$ in (1.10.1), the function $\overline{H}_{P,Q}^{M,N}$ reduces to the Fox's H-function.

1.11. A-FUNCTION

${}_1(a_j,\alpha_j)_n$ Represents the set of n pairs of parameters the A-function was defined by Gautam and Goyal [3] as

$$A_{p,q}^{m,n}\left[x\left|\begin{matrix}{}_1(a_j,\alpha_j)_p\\ {}_1(b_j,\beta_j)_q\end{matrix}\right.\right] = \frac{1}{2\pi i}\int_L f(s)x^s ds \tag{1.11.1}$$

Where
$$f(s) = \frac{\prod_{j=1}^{m}\Gamma(a_j+\alpha_j s)\prod_{j=1}^{n}\Gamma(1-b_j-\beta_j s)}{\prod_{j=m+1}^{p}\Gamma(1-a_j-\alpha_j s)\prod_{j=n+1}^{q}\Gamma(b_j+\beta_j s)} \tag{1.11.2}$$

The integral on the right hand side of (1.11.1) is convergent when $f > 0$ and $|\arg(ux)| < \frac{f\pi}{2}$, where

$$f = \operatorname{Re}(\sum_{j=1}^{m}\alpha_j - \sum_{j=m+1}^{p}\alpha_j + \sum_{j=1}^{n}\beta_j - \sum_{j=n+1}^{q}\beta_j$$

$$u = \prod_{j=1}^{p}\alpha_j^{\alpha_j}\prod_{j=1}^{q}\beta_j^{-\beta_j} \tag{1.11.3}$$

(1.11.1) reduces to H-function given by Fox the following relation

$$A_{p,q}^{n,m}\left[x\left|\begin{matrix}{}_1(1-a_j,\alpha_j)_p\\ {}_1(1-b_j,\beta_j)_q\end{matrix}\right.\right]=H_{p,q}^{m,n}\left[x\left|\begin{matrix}{}_1(a_j,\alpha_j)_p\\ {}_1(b_j,\beta_j)_q\end{matrix}\right.\right] \quad (1.11.4)$$

1.12. FRACTIONAL INTEGRAL OPERATORS

The definition of fractional integral of order α studied by Riemann-Liouville (1832, a, 76) as follows:

$$f_\alpha^+(a,x)=\frac{1}{\Gamma(\alpha)}\int_a^x f(t)(x-t)^{\alpha-1}dt \quad \text{(Right hand)} \quad (1.12.1)$$

$$f_\alpha^-(x,b)=\frac{1}{\Gamma(\alpha)}\int_x^b f(t)(t-x)^{\alpha-1}dt \quad \text{(Left hand)} \quad (1.12.2)$$

Where $a \le x \le b$, $\alpha > 0$ and is Γ a gamma function.

Weyl (1917) defined the fractional integral of order α in the form:

$$f_\alpha^+(-\infty,x)=\frac{1}{\Gamma(\alpha)}\int_{-\infty}^x f(t)(x-t)^{\alpha-1}dt \quad (1.12.3)$$

$$f_\alpha^-(x,+\infty)=\frac{1}{\Gamma(\alpha)}\int_x^{+\infty} f(t)(t-x)^{\alpha-1}dt \quad (1.12.4)$$

Erdelyi (1940) defined the fractional integral of order α as

$$I_x^\alpha f(x)=\frac{1}{\Gamma(\alpha)}\int_0^x (x-t)^{\alpha-1}f(t)dt,$$

$$I_x^0 f(x)=f(x) \quad (1.12.5)$$

$$K_x^\alpha f(x)=\frac{1}{\Gamma(\alpha)}\int_x^\infty (t-x)^{\alpha-1}f(t)dt,$$

$$K_x^0 f(x)=f(x) \quad (1.12.6)$$

Kober (1940) defined and studied the following fractional integrals of order α (using Erdelyi's notation):

$$\zeta\left[f(x)\right]=\zeta\left[\alpha,\beta,\gamma;m,\mu,\eta,a;f(x)\right]=\frac{\mu x^{-\eta-1}}{\Gamma(1-\alpha)}\int_0^x F(\alpha,\beta+m;\gamma;\frac{ax^\mu}{t^\mu})t^\eta f(t)dt$$

$$I_x^{\eta,\alpha}f(x)=x^{-\eta-\alpha}I_x^\alpha x^\eta f(x),\quad I_x^{\eta,0}f(x)=f(x) \quad (1.12.7)$$

$$K_x^{\eta,\alpha} f(x) = x K_x^{\alpha} x^{-\eta-\alpha} f(x), \quad K_x^{\eta,0} f(x) = f(x) \tag{1.12.8}$$

Saxena,(1967b) introduced and studied the operators associated with a hypergeometric associated with a hypergeometric function in the following form:

$$\zeta[f(x)] = \zeta[\alpha, \beta; \gamma; m; f(x)]$$

$$= \frac{x^{-\gamma-1}}{\Gamma(1-\alpha)} \int_0^x F(\alpha, \beta+m; \beta; t/x) t^{\gamma} f(t) dt \tag{1.12.9}$$

$$\Re[f(x)] = \Re[\alpha, \beta; \delta; m; f(x)]$$

$$= \frac{x^{\delta}}{\Gamma(1-\alpha)} \int_x^{\infty} F(\alpha, \beta+m; \beta; t/x) t^{-\delta-1} f(t) dt \tag{1.12.10}$$

Where $F(\alpha, \beta; \gamma; x)$ is the ordinary hypergeometric function, and $\alpha, \beta, \gamma, \delta$ are complex parameters, if $m = 0$, these operators reduce to the operators due to Kober (1940).

Kalla and Saxena (1969) generalized the operators (1.12.9) and (1.12.10) by means of the following equations:

$$\zeta[f(x)] = \zeta[\alpha, \beta, \gamma; m, \mu, \eta, a; f(x)]$$

$$= \frac{\mu x^{-\eta-1}}{\Gamma(1-\alpha)} \int_0^x F(\alpha, \beta+m; \gamma; \frac{ax^{\mu}}{t^{\mu}}) t^{\eta} f(t) dt \tag{1.12.11}$$

$$\Re[f(x)] = \Re[\alpha, \beta, \gamma; m, \mu, \delta, a; f(x)]$$

$$= \frac{\mu x^{\delta}}{\Gamma(1-\alpha)} \int_x^{\infty} F(\alpha, \beta+m; \gamma; \frac{ax^{\mu}}{t^{\mu}}) t^{-\delta-1} f(t) dt \tag{1.12.12}$$

Where $\alpha, \beta, \gamma, \eta, \delta$ and a are complex parameters.

Lowndes (1970) generalized the Kober operators as well as Hankel operators and also derived their inverses.

Saxena, (1966) obtained an inversion formula for the Verma transform by the application of fractional integration operators.

Fox (1963) studied the integral transform in the light of the theory of fractional integration.

Fox (1971, 72) has enumerated the application of L and L^{-1} operators defined below, in solving the integral equations.

$$x^{-\alpha-\beta}L^{-1}[t^{-\alpha}\{x^{\beta}f(x)\}] = I[\alpha,\beta;f(x)] \tag{1.12.13}$$

$$x^{1-\alpha-\beta}L^{-1}[t^{-\alpha}\{x^{\beta-1}f(\frac{1}{x})\}],(X=1/X)$$

$$= R[\alpha,\beta;f(x)] \tag{1.12.14}$$

Saxena and Kumbhat (1973) introduced a generalization of Kober operators, and further in 1973, they introduced two new fractional integration operators associated with generalized H-function, and also derived their important properties. In another paper (1973), they established some theorems connecting L, L^{-1} and fractional integration operators, which is an extension of the work of Fox (1972). Saxena and Modi (1980, 85) defined and studied the multidimensional fractional integration operators associated with hypergeometric functions. Gupta and Garg (1984) studied certain multidimensional fractional integral operators involving a general multivariable function in their karnel and also established a relationship between multidimensional fractional integral operators and multidimensional integral transforms. The operators involving multivariable H-function as kernal were defined by Banerji and Sethi (1978).

This subject has also been enriched by the researches of workers like Love (1967, 70), Saigo (1984), Srivastava et al. (1990), Hardy and Littlewood (1925), Gupta and Rajani (1988, 90), Kalla (1966, 71), Mathur and krishana (1977), Pathak and Pandey (1989), Lowndes (1985) and several others.

A detailed account of this subject can be found explicitly in the various monographs notably by Nishimoto (1982, 84, 87, 91), Ross (1975), Samko and Kilbas (1987), Mcbride and Roach (1985), Oldham and Spanier (1974) and many others.

LITERATURE REVIEW

Buschman , and Srivastava, (1990). Motivated by some further examples of the use of Feynman integrals which arise in perturbation calculations of the equilibrium properties of a magnetic model of phase transitions, a generalization of the familiar H function.

Saxena, Ram, (1990). A new theorem concerning the Whittaker transform of two variables .The result is derived by the application of two dimensional Erdelyi-Kober operators of Weyl type.

Goyal, Jain and Gaur, (1991). Derive three new and interesting composition formulas for a general class of fractional integral operators involving the product of a general class of polynomials, a general sequence of functions and a multivariable H-function.

Chandel,Agarwal,and Kumar,(1992). Make an application of an integral involving sine functions, exponential functions, the product of the Kampe de Feriet functions and the multivariable H-function of Srivastava and Panda to evaluate three Fourier series. Also evaluate a multiple integral involving the multivariable H-function of Srivastava and Panda and make its application to derive a multiple exponential Fourier series.

Goyal, Jain and Gaur, (1992). It is shown how these operators can be identified with elements of the algebra of functions having the Mellin convolution as the product. Inversion formulas and the relation of our operators with the generalized Hankel transforms are also established.

Raina,Kiryakova,Saigo,(1995). The generalized fractional integration operators defined by kiryakova are expressed in terms of the Laplace transform L and it's inverse L^{-1}. These decomposition results are established on L_p spaces and some examples are deduced as their special cases.

Srivastava,Hussain,(1995). Derive a number of key formulas for the fractional integration of the multivariable H-function . Each of the general Eulerian integral formulas are shown to yield interesting new results for various families of generalized hypergeometric functions of several variables. Some of these applications of the key formulas would provide potentially useful generalizations of known results in the theory of fractional calculus.

Rathie, (1997). A natural generalization of the familiar H -function of Fox namely the I -function is proposed. Convergence conditions, various series representations, elementary properties and special cases for the I -function have also been given.

Haubold,and Mathai,(2000). The solution of a simple kinetic equation of the type used for the computation of the change of the chemical composition in stars like the Sun. Starting from the standard form of the kinetic equation it is generalized to a fractional kinetic equation and its solutions in terms of H-functions are obtained. The role of thermonuclear functions, which are also represented in terms of G- and H-functions, in such a fractional kinetic equation is emphasized. Results contained in this paper are related to recent investigations of possible astrophysical solutions of the solar neutrino problem.

Chaurasia,Singhal,(2004). Eulerian integral and a main theorem based upon the fractional integral operator associated with generalized polynomials given by Srivastava and H- function of several complex variables given by Srivastava and Panda which provide unification and extension of numerous results in the theory of fractional calculus of special functions in one and more variables.

Mathai,Saxena,Haubold,(2004). An analytic proof of the integrals for astrophysical thermonuclear functions which are derived on the basis of Boltzmann-Gibbs statistical mechanics. Among the four different cases of astrophysical thermonuclear functions, those with a depleted high-energy tail and a cut-off at high energies find a natural interpretation in q-statistics.

Suthar, Saxena and Ram, (2004). Derive the images of the product of the H-function, in which its argument contains a factor $zr^{k}(t^{k}+c^{\mu})^{-\rho}$, and a general class of polynomials defined by Srivastava under the multiple Erdelyi-Kober operators due to Galue et al. The results obtained are general in character and includes as special cases.

Chaurasia, and Srivastava,(2006). Establish a relation between the Two dimensional $\overline{H}$ transform involving a general polynomials and the weyl type two dimensional Saigo operator of fractional integral calculus.

Gupta, (2007). Introduce and study a new pair of fractional integral operators involving the product of H -function, Fox H-function and the general sequence of functions as their kernels.

Purohit, Yadav, Kalla,(2008). Certain expansion formulae for a basic analogue of the Fox's H-function have been derived by the applications of the q-Leibniz rule for the Weyl type q-derivatives of a product of two functions. Expansion formulae involving a basic analogue of Meijer's G-function and MacRobert's E-function have been derived as special cases of the main results.

Saxena, Ram, Chandak and Kalla , (2008). Evaluate two unified fractional integrals involving the product of Fox-Wright generalized hypergeometric function, Appell function, and a general class of multivariable polynomials. These integrals are further applied in proving two theorems on Saigo – Maeda operators of fractional integration.

Yadav, Purohit ,Kalla,(2008). Fractional q-integral operators of generalized Weyl type, involving generalized basic hypergeometric functions and a basic analogue of Fox's H- function have been investigated. A number of integrals involving various q-functions have been evaluated as applications of the main results.

Bhatter, Shekhawat, (2010). Multidimensional fractional integral operators whose kernels involve the product of multivariable polynomial H -function. **Chaurasia, Saxena,(2010).** Evaluate certain triple integral relations involving H-function and the multivariable H-function.

Sharma, and Singh. (2010). An attempt has been made to derive finite summation formulae for the H-function introduced by Srivastava and Panda. Since the multivariable H- Function includes a large number of a special function of one and more variables as its particular cases.

Singh, (2010). Establish a relation between the double Laplace transform and the double Hankel transform. A double Mellin-Hankel transform of the product of H-functions of one and two variables is then obtained.

Yadav and Purohit, (2010). Introduce two generalized operators of fractional q-integration, which may be regarded as extensions of Riemann-Liouville, Weyl and Kober fractional q-integral operators. Certain interesting connection theorems involving these operators and q-Mellin transform are also discussed.

Bhatter,Shekhawat,(2011). Establish a generalized fractional derivative operator formula involving a general multivariable polynomial and multivariable H- function.

Garg,Chanchlani,(2011). Right and left sided Kober fractional q-derivative operators and show that these derivative operators are left inverse operators of Kober fractional q-integral operators. Obtain the images of generalized basic hypergeometric function and basic analogue of Fox H-function under these operators. Also deduce several interesting results involving q-analogues of some classical functions as special cases are findings.

Lin, Liu, and Srivastava, (2011). Investigate several general families of hypergeometric polynomials and their associated multiple integral representations. Each of the integral representations, which are derived in this paper, may be viewed also as a linearization relationship for the product of two different members.

Saha,Arora,(2011). Establish a theorem on Weyl fractional derivatives of the product of hypergeometric function and the H-function. Certain special cases of theorem have also been discussed.

Satyanarayana,Kumar,(2011). Establish the integrals involving the product of two general classes of polynomial, H-function of one variable and H-function of 'r' variables. These integrals are unified in nature and we can derive from them by a large number of integrals involving simpler functions and polynomials as their particular cases.

CHAPTER 2

SECTION 1

THE RELATION BETWEEN

DOUBLE LAPLACE TRANSFORM

AND

DOUBLE HANKEL TRANSFORM

WITH

APPLICATIONS

The object of this chapter is to establish a relation between the double Laplace transform and the double Hankel transform. A double Laplace-Hankel transform of the product of H-functions of one and two variables is then obtained. Application of our main result, summation formula and some interesting special cases has also been discussed.

1. Introduction:

If $F(p_1, p_2)$ is theDouble Laplace transform of $f(x,y)$, then

$$F(p_1,p_2)=\int_0^\infty\int_0^\infty e^{-p_1x-p_2y}f(x,y)dxdy;\quad \mathrm{Re}(p_1)>0,\mathrm{Re}(p_2)>0 \tag{2.1.1}$$

If $H(p_1, p_2)$ is the Double Hankel transform of $f(x,y)$, then

$$H(p_1,p_2)=\int_0^\infty\int_0^\infty xJ_\nu(p_1x)yJ_\mu(p_2y)f(x,y)dxdy \tag{2.1.2}$$

Provided that $xJ_\nu(p_1x)>0, yJ_\mu(p_2y)>0$.

The following formula is required in the proof:

$$\int_0^\infty\int_0^\infty x^{s-1}y^{t-1}H\left[ax^\lambda,by^\mu\right]dxdy=\frac{a^{-s/\lambda}b^{-t/\mu}}{\lambda\mu}\phi\left(-\frac{s}{\lambda},-\frac{t}{\mu}\right)\theta_2\left(-\frac{s}{\lambda}\right)\theta_3\left(-\frac{t}{\mu}\right) \tag{2.1.3}$$

H[x] represents the H-function of Fox (1961).

The H-function of two variables (Mittal and Gupta (1972), p.172) using the following notation, which is due essentially to Srivastava and Panda ((1976 a), p.266, eq. (1.5) et seq.)is defined and represented as:

$$H[x,y]=H\left[\begin{smallmatrix}x\\y\end{smallmatrix}\right]=H^{0,n_1:m_2,n_2:m_3,n_3}_{p_1,q_1:p_2,q_2:p_3,q_3}\left[\begin{matrix}x\\y\end{matrix}\middle|\begin{matrix}(a_j;\alpha_j,A_j)_{1,p_1}:(c_j,\gamma_j)_{1,p_2},(e_j,E_j)_{1,p_3}\\(b_j;\beta_j,B_j)_{1,q_1}:(d_j,\delta_j)_{1,q_2},(f_j,F_j)_{1,q_3}\end{matrix}\right]$$

$$=-\frac{1}{4\pi^2}\int_{L_1}\int_{L_2}\phi(\xi,\eta)\theta_2(\xi)\theta_3(\eta)x^\xi y^\eta d\xi\, d\eta \tag{2.1.4}$$

Where

$$\phi(\xi,\eta)=\frac{\prod_{j=1}^{n_1}\Gamma(1-a_j+\alpha_j\xi+A_j\eta)}{\prod_{j=n_1+1}^{p_1}\Gamma(a_j-\alpha_j\xi-A_j\eta)\prod_{j=1}^{q_1}\Gamma(1-b_j+\beta_j\xi+B_j\eta)} \tag{2.1.5}$$

$$\theta_2(\xi)=\frac{\prod_{j=1}^{n_2}\Gamma(1-c_j+\gamma_j\xi)\prod_{j=1}^{m_2}\Gamma(d_j-\delta_j\xi)}{\prod_{j=n_2+1}^{p_2}\Gamma(c_j-\gamma_j\xi)\prod_{j=m_2+1}^{q_2}\Gamma(1-d_j+\delta_j\xi)} \tag{2.1.6}$$

$$\theta_3(\eta)=\frac{\prod_{j=1}^{n_3}\Gamma(1-e_j+E_j\eta)\prod_{j=1}^{m_3}\Gamma(f_j-F_j\eta)}{\prod_{j=n_3+1}^{p_3}\Gamma(e_j-E_j\eta)\prod_{j=m_3+1}^{q_3}\Gamma(1-f_j+F_j\eta)} \tag{2.1.7}$$

2. Main Result:

Theorem: If $F(p_1,p_2)$ is the Laplace transform, then

$$F(p_1,p_2)=\frac{(p_1)^{-\nu-s_1}(-p_2)^{-\mu-s_2}}{2^{-\nu-\mu-2(s_1+s_2)}}\Gamma(\nu+s_1+1)\Gamma(\mu+s_2+1)$$
$$\int_0^\infty\int_0^\infty t_1^{-s_1-\nu}t_2^{-s_2-\mu}J_\nu(p_1t_1)J_\mu(p_2t_2)f(t_1,t_2)\,dt_1\,dt_2 \tag{2.1.8}$$

Proof: From (2.1.1), $F(p_1,p_2)=\int_0^\infty\int_0^\infty e^{-p_1t_1-p_2t_2}f(t_1,t_2)\,dt_1\,dt_2$

$$=\sum_{s_1=0}^{\infty}\sum_{s_2=0}^{\infty}\frac{(-p_1)^{s_1}}{s_1!}\frac{(-p_2)^{s_2}}{s_2!}\int_0^\infty\int_0^\infty t_1^{s_1}t_2^{s_2}f(t_1,t_2)\,dt_1\,dt_2$$

=

$$\frac{(p_1)^{-\nu-s_1}(-p_2)^{-\mu-s_2}}{2^{-\nu-\mu-2(s_1+s_2)}}\Gamma(\nu+s_1+1)\Gamma(\mu+s_2+1)\int_0^\infty\int_0^\infty t_1^{-s_1-\nu}t_2^{-s_2-\mu}J_\nu(p_1t_1)J_\mu(p_2t_2)f(t_1,t_2)\,dt_1\,dt_2$$

3. A Double Hankel Transform:

$$\int_0^\infty\int_0^\infty xJ_\nu(p_1x)yJ_\mu(p_2y)H_{p,q}^{m,n}\left[cx^\lambda y^\delta\left|\begin{matrix}{}_1(g_j,G_j)_p\\ {}_1(h_j,H_j)_q\end{matrix}\right.\right]H\left[ax^\gamma,by^\eta\right]dxdy$$

$$= \sum_{r_1=0}^{\infty}\sum_{r_2=0}^{\infty} \frac{(-1)^{r_1}(-1)^{r_2} p_1^{\nu+2r_1} p_2^{\mu+2r_2}}{r_1!r_2!\Gamma(\nu+r_1+1)\Gamma(\mu+r_2+1)\,2^{\nu+\mu+2(r_1+r_2)}} a^{-(\nu+2r_1+2)/\gamma} b^{-(\mu+2r_2+2)/\eta} (\gamma\eta)^{-1}$$

$$H^{m+n_2+n_3,n+m_2+m_3}_{p+q_1+q_2+q_3,q+p_1+p_2+p_3}\left[ca^{-\lambda/\gamma} b^{-\delta/\eta} \middle| \begin{matrix} {}_1(g_j,G_j)_n,{}_1(1-d_j-\left(\frac{\nu+2r_1+2}{\gamma}\right)D_j;\frac{\lambda}{\gamma}D_j)_{m_2} \\ {}_1(h_j,H_j)_m,{}_1(1-c_j-\left(\frac{\nu+2r_1+2}{\gamma}\right)C_j;\frac{\lambda}{\gamma}C_j)_{m_2} \end{matrix} \right.$$

$$\begin{matrix} {}_1\left(1-f_j-\left(\frac{\mu+2r_2+2}{\eta}\right)F_j,\frac{\delta}{\eta}F_j\right)_{p_3},{}_{n+1}(g_j,G_j)_p,{}_{n_2+1}\left(1-d_j-\left(\frac{\nu+2r_1+2}{\gamma}\right)D_j,\frac{\lambda}{\gamma}D_j\right)_{q_2} \\ {}_1\left(1-e_j-\left(\frac{\mu+2r_2+2}{\eta}\right)E_j,\frac{\delta}{\eta}E_j\right)_{p_3},{}_{m+1}(h_j,H_j)_q,{}_{n_2+1}\left(1-c_j-\left(\frac{\nu+2r_1+2}{\gamma}\right)C_j,\frac{\lambda}{\gamma}C_j\right)_{p_2} \end{matrix}$$

$$\left. \begin{matrix} {}_1\left(1-b_j-\left(\frac{\nu+2r_1+2}{\gamma}\right)\beta_j-\left(\frac{\mu+2r_2+2}{\eta}\right)B_j+\frac{\lambda}{\gamma}\beta_j+\frac{\delta}{\eta}B_j\right)_{q_1} \\ {}_1\left(1-a_j-\left(\frac{\nu+2r_1+2}{\gamma}\right)\alpha_j-\left(\frac{\mu+2r_2+2}{\eta}\right)A_j+\frac{\lambda}{\gamma}\alpha_j+\frac{\delta}{\eta}A_j\right)_{p_1} \end{matrix} \right] \tag{2.1.9}$$

Provided,

$$\lambda,\delta > 0; \eta > 0; |\arg c| < \frac{1}{2}\Delta\pi, \Delta > 0$$

Where

$$\Delta = \sum_{j=1}^{m} H_j - \sum_{j=m+1}^{q} H_j + \sum_{j=1}^{n} G_j - \sum_{j=n+1}^{p} G_j$$

$$\mathrm{Re}[(\nu+2r_1+2+\gamma(d_i/D_i)+\lambda(h_j/H_j)] > 0; i=1,\dots,m_2; j=1,\dots,m$$

$$\mathrm{Re}[(\mu+2r_2+2+\eta(f_i/F_i)+\delta(h_j/H_j)] > 0; i=1,\dots,m_3; j=1,\dots,m$$

$$\mathrm{Re}\left[\nu+2r_1+2-\gamma\left(\frac{1-c_i}{C_i}\right)-\lambda\left(\frac{1-g_j}{G_j}\right)\right] < 0; i=1,..,n_2; j=1,\dots,n$$

Proof: To prove (2.1.9), Expand Bessel function in series form and substitute the Mellin-Bernes contour integral for $H[cx^{\lambda}y^{\delta}]$on the left hand side then interchange the order of contour integral and the (x,y)-integrals. Finally we arrive at our result on evaluating the (x,y) integral by using the result (2.1.3).

4. Application of the Main Result:

If $f(t_1,t_2)=t_1^{\nu+2r_1+2}t_2^{\mu+2r_2+2}J_\nu(p_1t_1)J_\mu(p_2t_2)H_{p,q}^{m,n}\left[ct_1^\lambda t_2^\delta\left|\begin{smallmatrix}{}_1(g_j,G_j)_p\\ {}_1(h_j,H_j)_q\end{smallmatrix}\right.\right]H\left[at^\gamma,bt^\eta\right]$;

(2.1.8) becomes the double Laplace-Hankel transform of the product of H-functions of one and two variables and take the following form:

$$\int_0^\infty\int_0^\infty e^{-P_1t_1-P_2t_2}t_1^{\nu+2r_1+2}t_2^{\mu+2r_2+2}J_\nu(p_1t_1)J_\mu(p_2t_2)H_{p,q}^{m,n}\left[ct_1^\lambda t_2^\delta\left|\begin{smallmatrix}{}_1(g_j,G_j)_p\\ {}_1(h_j,H_j)_q\end{smallmatrix}\right.\right]H\left[at_1^\gamma,bt_2^\eta\right]dt_1dt_2=$$

$$\sum_{s_1=0}^\infty\sum_{s_2=0}^\infty\sum_{r_1=0}^\infty\sum_{r_2=0}^\infty\frac{(-1)^{s_1}(-1)^{s_2}(-1)^{r_1}(-1)^{r_2}p_1^{\nu+s_1+2r_1}p_2^{\mu+s_2+2r_2}}{s_1!s_2!r_1!r_2!\Gamma(\nu+r_1+1)\Gamma(\mu+r_2+1)2^{\nu+\mu+2(r_1+r_2)}}a^{-(\nu+2r_1+2)/\gamma}b^{-(\mu+2r_2+2)/\eta}(\gamma\eta)^{-1}$$

$$H_{p+q_1+q_2+q_3,q+p_1+p_2+p_3}^{m+n_2+n_3,n+m_2+m_3}\left[ca^{-\lambda/\gamma}b^{-\delta/\eta}\left|\begin{matrix}{}_1(g_j,G_j)_n,{}_1(1-d_j-\left(\frac{\nu+2r_1+s_1+2}{\gamma}\right)D_j;\frac{\lambda}{\gamma}D_j)_{m_2}\\ {}_1(h_j,H_j)_m,{}_1(1-c_j-\left(\frac{\nu+2r_1+s_1+2}{\gamma}\right)C_j;\frac{\lambda}{\gamma}C_j)_{n_2}\end{matrix}\right.\right.$$

$$\begin{matrix}{}_1\left(1-f_j-\left(\frac{\mu+2r_2+s_2+2}{\eta}\right)F_j,\frac{\delta}{\eta}F_j\right)_{p_3},{}_{n+1}(g_j,G_j)_p,{}_{m_2+1}\left(1-d_j-\left(\frac{\nu+2r_1+s_1+2}{\gamma}\right)D_j,\frac{\lambda}{\gamma}D_j\right)_{q_2}\\ {}_1\left(1-e_j-\left(\frac{\mu+2r_2+s_2+2}{\eta}\right)E_j,\frac{\delta}{\eta}E_j\right)_{p_3},{}_{m+1}(h_j,H_j)_q,{}_{n_2+1}\left(1-c_j-\left(\frac{\nu+2r_1+s_1+2}{\gamma}\right)C_j,\frac{\lambda}{\gamma}C_j\right)_{p_2}\end{matrix}$$

$$\left.\begin{matrix}{}_1\left(1-b_j-\left(\frac{\nu+2r_1+s_1+2}{\gamma}\right)\beta_j-\left(\frac{\mu+2r_2+s_2+2}{\eta}\right)B_j,\frac{\lambda}{\gamma}\beta_j+\frac{\delta}{\eta}B_j\right)_{q_1}\\ {}_1\left(1-a_j-\left(\frac{\nu+2r_1+s_1+2}{\gamma}\right)\alpha_j-\left(\frac{\mu+2r_2+s_2+2}{\eta}\right)A_j,\frac{\lambda}{\gamma}\alpha_j+\frac{\delta}{\eta}A_j\right)_{p_1}\end{matrix}\right] \quad (2.1.10)$$

Provided the conditions are same as that of (2.1.9) with $Re(p_1)>0$, $Re(p_2)>0$.

Proof: In(2.1.8)put

$$f(t_1,t_2)=t_1^{\nu+2r_1+2}t_2^{\mu+2r_2+2}J_\nu(p_1t_1)J_\mu(p_2t_2)H_{p,q}^{m,n}\left[ct_1^\lambda t_2^\delta\left|\begin{smallmatrix}{}_1(g_j,G_j)_p\\ {}_1(h_j,H_j)_q\end{smallmatrix}\right.\right]H\left[at^\gamma bt^\eta\right]$$

and use (2.1.9) to get,

$$H(p_1,p_2)=\sum_{r_1=0}^\infty\sum_{r_2=0}^\infty\frac{(-1)^{r_1}(-1)^{r_2}p_1^{\nu+2r_1}p_2^{\mu+2r_2}}{r_1!r_2!\Gamma(\nu+r_1+1)\Gamma(\mu+r_2+1)2^{\nu+\mu+2(r_1+r_2)}}a^{-(\nu+2r_1+1)/\gamma}b^{-(\mu+2r_2+1)/\eta}(\gamma\eta)^{-1}$$

$$H^{m+n_2+n_3,n+m_2+m_3}_{p+q_1+q_2+q_3,q+p_1+p_2+p_3}\left[ca^{-\lambda/\gamma}b^{-\delta/\eta}\left|\begin{matrix}{}_1(g_j,G_j)_n,{}_1(1-d_j-\left(\frac{\nu+2r_1+1}{\gamma}\right)D_j;\frac{\lambda}{\gamma}D_j)_{m_2}\\{}_1(h_j,H_j)_m,{}_1(1-c_j-\left(\frac{\nu+2r_1+1}{\gamma}\right)C_j;\frac{\lambda}{\gamma}C_j)_{n_2}\end{matrix}\right.\right.$$

$$\begin{matrix}{}_1\left(1-f_j-\left(\frac{\mu+2r_2+1}{\eta}\right)F_j,\frac{\delta}{\eta}F_j\right)_{p_3},{}_{n+1}(g_j,G_j)_p,{}_{n_2+1}\left(1-d_j-\left(\frac{\nu+2r_1+1}{\gamma}\right)D_j,\frac{\lambda}{\gamma}D_j\right)_{q_2}\\{}_1\left(1-e_j-\left(\frac{\mu+2r_2+1}{\eta}\right)E_j,\frac{\delta}{\eta}E_j\right)_{p_3},{}_{m+1}(h_j,H_j)_q,{}_{n_2+1}\left(1-c_j-\left(\frac{\nu+2r_1+1}{\gamma}\right)C_j,\frac{\lambda}{\gamma}C_j\right)_{p_2}\end{matrix}$$

$$\left.\begin{matrix}{}_1\left(1-b_j-\left(\frac{\nu+2r_1+1}{\gamma}\right)\beta_j-\left(\frac{\mu+2r_2+1}{\eta}\right)B_j+\frac{\lambda}{\gamma}\beta_j+\frac{\delta}{\eta}B_j\right)_{q_1}\\{}_1\left(1-a_j-\left(\frac{\nu+2r_1+1}{\gamma}\right)\alpha_j-\left(\frac{\mu+2r_2+1}{\eta}\right)A_j+\frac{\lambda}{\gamma}\alpha_j+\frac{\delta}{\eta}A_j\right)_{p_1}\end{matrix}\right]$$

Hence $F(p_1, p_2)$=the right hand side of (2.1.10).

5. Summation formula:

$$\int_0^\infty\int_0^\infty e^{-p_1t_1-p_2t_2}J_\nu(p_1t_1)J_\mu(p_2t_2)H^{m,n}_{p,q}\left[ct_1^\lambda t_2^\delta\left|\begin{matrix}{}_1(g_j,G_j)_p\\{}_1(h_j,H_j)_q\end{matrix}\right.\right]H\left[at_1^\gamma,bt_2^\eta\right]dt_1dt_2=$$

$$\sum_{s_1=0}^\infty\sum_{s_2=0}^\infty\sum_{r_1=0}^\infty\sum_{r_2=0}^\infty\frac{(-1)^{s_1}(-1)^{s_2}(-1)^{r_1}(-1)^{r_2}p_1^{\nu+s_1+2r_1}p_2^{\mu+s_2+2r_2}}{s_1!s_2!r_1!r_2!\Gamma(\nu+r_1+1)\Gamma(\mu+r_2+1)2^{\nu+\mu+2(r_1+r_2)}}a^{-(\nu+2r_1+2)/\gamma}b^{-(\mu+2r_2+2)/\eta}(\gamma\eta)^{-1}$$

$$H^{m+n_2+n_3,n+m_2+m_3}_{p+q_1+q_2+q_3,q+p_1+p_2+p_3}\left[ca^{-\lambda/\gamma}b^{-\delta/\eta}\left|\begin{matrix}{}_1(g_j,G_j)_n,{}_1(1-d_j-\left(\frac{\nu+2r_1+s_1+2}{\gamma}\right)D_j;\frac{\lambda}{\gamma}D_j)_{m_2}\\{}_1(h_j,H_j)_m,{}_1(1-c_j-\left(\frac{\nu+2r_1+s_1+2}{\gamma}\right)C_j;\frac{\lambda}{\gamma}C_j)_{n_2}\end{matrix}\right.\right.$$

$$\begin{matrix}{}_1\left(1-f_j-\left(\frac{\mu+2r_2+s_2+2}{\eta}\right)F_j,\frac{\delta}{\eta}F_j\right)_{p_3},{}_{n+1}(g_j,G_j)_p,{}_{n_2+1}\left(1-d_j-\left(\frac{\nu+2r_1+s_1+2}{\gamma}\right)D_j,\frac{\lambda}{\gamma}D_j\right)_{q_2}\\{}_1\left(1-e_j-\left(\frac{\mu+2r_2+s_2+2}{\eta}\right)E_j,\frac{\delta}{\eta}E_j\right)_{p_3},{}_{m+1}(h_j,H_j)_q,{}_{n_2+1}\left(1-c_j-\left(\frac{\nu+2r_1+s_1+2}{\gamma}\right)C_j,\frac{\lambda}{\gamma}C_j\right)_{p_2}\end{matrix}$$

$$\left.\begin{matrix}{}_1\left(1-b_j-\left(\frac{\nu+2r_1+1}{\gamma}\right)\beta_j-\left(\frac{\mu+2r_2+1}{\eta}\right)B_j+\frac{\lambda}{\gamma}\beta_j+\frac{\delta}{\eta}B_j\right)_{q_1}\\{}_1\left(1-a_j-\left(\frac{\nu+2r_1+1}{\gamma}\right)\alpha_j-\left(\frac{\mu+2r_2+1}{\eta}\right)A_j+\frac{\lambda}{\gamma}\alpha_j+\frac{\delta}{\eta}A_j\right)_{p_1}\end{matrix}\right] \quad (2.1.11)$$

Evaluating the left hand side of (2.1.11), using (Srivastava, Gupta and Goyal (1982) eq.(8.5.6),p.150), the following summation formula is obtained:

$$\sum_{r_1=0}^{\infty}\sum_{r_2=0}^{\infty}\frac{(-1)^{r_1}(-1)^{r_2}p_1^{-2}p_2^{-2}}{r_1!r_2!\Gamma(\nu+r_1+1)\Gamma(\mu+r_2+1)2^{\nu+\mu+2(r_1+r_2)}}a^{-\xi}b^{-\eta}$$

$$H^{m+n_2+n_3,n+2+m_2+m_3}_{p+2+q_1+q_2+q_3,q+p_1+p_2+p_3}\left[cp_1^{-\delta}p_2^{-\eta}ab\left|\begin{matrix}{}_1(g_j,G_j)_n,(-\nu-2r_1,\lambda),(-\mu-2r_2,\delta),{}_1(1-d_j;\delta D_j)_{m_2}\\ {}_1(h_j,H_j)_m,{}_1(1-c_j;\gamma C_j)_{n_2}\end{matrix}\right.\right.$$

$$\left.\begin{matrix}{}_1(1-f_j,\eta F_j)_{q_3},{}_{n+1}(g_j,G_j)_p,{}_{m_2+1}(1-d_j,\delta D_j)_{q_2},{}_1(1-b_j+\xi\beta_j+\eta B_j)_{q_1}\\ {}_1(1-e_j,\eta E_j)_{p_3},{}_{m+1}(h_j,H_j)_q,{}_{n_2+1}(1-c_j,\gamma C_j)_{p_2},{}_1(1-a_j+\xi\alpha_j+\eta A_j)_{p_1}\end{matrix}\right]$$

$$=$$

$$\sum_{s_1=0}^{\infty}\sum_{s_2=0}^{\infty}\sum_{r_1=0}^{\infty}\sum_{r_2=0}^{\infty}\frac{(-1)^{s_1}(-1)^{s_2}(-1)^{r_1}(-1)^{r_2}P_1^{\nu+s_1+2r_1}P_2^{\mu+s_2+2r_2}}{s_1!s_2!r_1!r_2!\Gamma(\nu+r_1+1)\Gamma(\mu+r_2+1)2^{\nu+\mu+2(r_1+r_2)}}a^{-(\nu+2r_1+2)/\gamma}b^{-(\mu+2r_2+2)/\eta}(\gamma\eta)^{-1}$$

$$H^{m+n_2+n_3,n+m_2+m_3}_{p+q_1+q_2+q_3,q+p_1+p_2+p_3}\left[ca^{-\lambda/\gamma}b^{-\delta/\eta}\left|\begin{matrix}{}_1(g_j,G_j)_n,{}_1\left(1-d_j-\left(\frac{\nu+2r_1+s_1+2}{\gamma}\right)D_j;\frac{\lambda}{\gamma}D_j\right)_{m_2}\\ {}_1(h_j,H_j)_m,{}_1\left(1-c_j-\left(\frac{\nu+2r_1+s_1+2}{\gamma}\right)C_j;\frac{\lambda}{\gamma}C_j\right)_{n_2}\end{matrix}\right.\right.$$

$$\begin{matrix}{}_1\left(1-f_j-\left(\frac{\mu+2r_2+s_2+2}{\eta}\right)F_j,\frac{\delta}{\eta}F_j\right)_{p_3},{}_{n+1}(g_j,G_j)_p,{}_{m_2+1}\left(1-d_j-\left(\frac{\nu+2r_1+s_1+2}{\gamma}\right)D_j,\frac{\lambda}{\gamma}D_j\right)_{q_2}\\ {}_1\left(1-e_j-\left(\frac{\mu+2r_2+s_2+2}{\eta}\right)E_j,\frac{\delta}{\eta}E_j\right)_{p_3},{}_{m+1}(h_j,H_j)_q,{}_{n_2+1}\left(1-c_j-\left(\frac{\nu+2r_1+s_1+2}{\gamma}\right)C_j,\frac{\lambda}{\gamma}C_j\right)_{p_2}\end{matrix}$$

$$\left.\begin{matrix}{}_1\left(1-b_j-\left(\frac{\nu+2r_1+1}{\gamma}\right)\beta_j-\left(\frac{\mu+2r_2+1}{\eta}\right)B_j,+\frac{\lambda}{\gamma}\beta_j+\frac{\delta}{\eta}B_j\right)_{q_1}\\ {}_1\left(1-a_j-\left(\frac{\nu+2r_1+1}{\gamma}\right)\alpha_j-\left(\frac{\mu+2r_2+1}{\eta}\right)A_j,+\frac{\lambda}{\gamma}\alpha_j+\frac{\delta}{\eta}A_j\right)_{p_1}\end{matrix}\right]$$

Provided the conditions are same as that of (2.1.9) with $Re(p_1)>0$, $Re(p_2)>0$.

6. Special Cases:

(1) In (2.1.10) let $P_1\to 0$, $P_2\to 0$ and $\gamma=\eta=1$, to get the following result:

$$\int_0^{\infty}\int_0^{\infty}t_1^{\nu+2r_1+2}t_2^{\mu+2r_2+2}J_\nu(P_1t_1)J_\mu(P_2t_2)H^{m,n}_{p,q}\left[ct_1^{\lambda}t_2^{\delta}\left|\begin{matrix}{}_1(g_j,G_j)_p\\ {}_1(h_j,H_j)_q\end{matrix}\right.\right]H\left[at_1,bt_2\right]dt_1dt_2=$$

$$\sum_{r_1=0}^{\infty}\sum_{r_2=0}^{\infty}\frac{(-1)^{r_1}(-1)^{r_2}}{r_1!r_2!\Gamma(\nu+r_1+1)\Gamma(\mu+r_2+1)2^{\nu+\mu+2(r_1+r_2)}}a^{-(\nu+2r_1+2)}b^{-(\mu+2r_2+2)}$$

$$H^{m+n_2+n_3,n+m_2+m_3}_{p+q_1+q_2+q_3,q+p_1+p_2+p_3}\left[ca^{-\lambda}b^{-\delta}\middle|\begin{matrix} {}_1(g_j,G_j)_n,{}_1(1-d_j-(\nu+2r_1+2)D_j;\lambda D_j)_{m_2} \\ {}_1(h_j,H_j)_m,{}_1(1-c_j-(\nu+2r_1+2)C_j;\lambda C_j)_{m_2}\end{matrix}\right.$$

$$\left.\begin{matrix} {}_1\left(1-f_j-(\mu+2r_2+2)F_j,\delta F_j\right)_{p_3},{}_{m+1}(g_j,G_j)_p,{}_{m_2+1}\left(1-d_j-(\nu+2r_1+2)D_j,\lambda D_j\right)_{q_2}\ {}_1\left(1-b_j-(\nu+2r_1+2)\beta_j-(\mu+2r_2+2)B_j+\lambda\beta_j+\delta B_j\right)_{q_1} \\ {}_1\left(1-e_j-(\mu+2r_2+2)E_j,\delta E_j\right)_{p_3},{}_{m+1}(h_j,H_j)_q,{}_{m_2+1}\left(1-c_j-(\nu+2r_1+2)C_j,\lambda C_j\right)_{p_2}\ {}_1\left(1-a_j-(\nu+2r_1+2)\alpha_j-(\mu+2r_2+2)A_j+\lambda\alpha_j+\delta A_j\right)_{p_1}\end{matrix}\right]$$

(2.1.12)

(2) In (2.1.10) take $n_1= p_1=q_1=0$, to get the double Laplace-Hankel transform of the product of three single H-functions of Fox as:

$$\int_0^\infty\int_0^\infty e^{-p_1t_1-p_2t_2}t_1^{\nu+2r_1+2}t_2^{\mu+2r_2+2}J_\nu(p_1t_1)J_\mu(p_2t_2)H^{m,n}_{p,q}\left[ct_1^\lambda t_2^\delta\middle|\begin{matrix}{}_1(g_j,G_j)_p\\{}_1(h_j,H_j)_q\end{matrix}\right]$$

$$H^{m_2,n_2}_{p_2,q_2}\left[ct_1^\gamma\middle|\begin{matrix}{}_1(c_j,C_j)_{p_2}\\{}_1(d_j,D_j)_{q_2}\end{matrix}\right]H^{m_3,n_3}_{p_3,q_3}\left[ct_2^\eta\middle|\begin{matrix}{}_1(e_j,E_j)_{p_3}\\{}_1(f_j,F_j)_{q_3}\end{matrix}\right]dt_1dt_2=$$

$$\sum_{s_1=0}^\infty\sum_{s_2=0}^\infty\sum_{r_1=0}^\infty\sum_{r_2=0}^\infty\frac{(-1)^{s_1}(-1)^{s_2}(-1)^{r_1}(-1)^{r_2}p_1^{\nu+s_1+2r_1}p_2^{\mu+s_2+2r_2}}{s_1!s_2!r_1!r_2!\Gamma(\nu+r_1+1)\Gamma(\mu+r_2+1)2^{\nu+\mu+2(r_1+r_2)}}a^{-(\nu+2r_1+2)/\gamma}b^{-(\mu+2r_2+2)/\eta}(\gamma\eta)^{-1}$$

$$H^{m+n_2+n_3,n+m_2+m_3}_{p+q_1+q_2+q_3,q+p_1+p_2+p_3}\left[ca^{-\lambda/\gamma}b^{-\delta/\eta}\middle|\begin{matrix}{}_1(g_j,G_j)_n,{}_1(1-d_j-\left(\frac{\nu+2r_1+s_1+2}{\gamma}\right)D_j;\frac{\lambda}{\gamma}D_j)_{m_2}\\{}_1(h_j,H_j)_m,{}_1(1-c_j-\left(\frac{\nu+2r_1+s_1+2}{\gamma}\right)C_j;\frac{\lambda}{\gamma}C_j)_{m_2}\end{matrix}\right.$$

$$\left.\begin{matrix}{}_1\left(1-f_j-\left(\frac{\mu+2r_2+s_2+2}{\eta}\right)F_j,\frac{\delta}{\eta}F_j\right)_{p_3},{}_{m+1}(g_j,G_j)_p,{}_{m_2+1}\left(1-d_j-\left(\frac{\nu+2r_1+s_1+2}{\gamma}\right)D_j,\frac{\lambda}{\gamma}D_j\right)_{q_2}\\{}_1\left(1-e_j-\left(\frac{\mu+2r_2+s_2+2}{\eta}\right)E_j,\frac{\delta}{\eta}E_j\right)_{p_3},{}_{m+1}(h_j,H_j)_q,{}_{m_2+1}\left(1-c_j-\left(\frac{\nu+2r_1+s_1+2}{\gamma}\right)C_j,\frac{\lambda}{\gamma}C_j\right)_{p_2}\end{matrix}\right]$$

(2.1.13)

Provided the conditions are same as that of (2.1.9) with $p_1=q_1=0$; $Re(p_1)>0$, $Re(p_2)>0$.

SECTION 2

RELATIONSHIP BETWEEN DOUBLE LAPLACE TRANSFORM AND DOUBLE MELLIN TRANSFORM IN TERMS OF GENERALIZED HYPERGEOMETRIC FUNCTION WITH APPLICATIONS

The object of this chapter is to establish a relation between the double Laplace transform and the double Mellin transform. A double Laplace-Mellin transform of the product of H-functions of one and two variables is then obtained. Application, summation formula and some interesting special cases have also been discussed.

1. Introduction:

If $F(p_1, p_2)$ is the Double Laplace transform of $f(x, y)$, then

$$F(p_1, p_2) = \int_0^\infty \int_0^\infty e^{-p_1 x - p_2 y} f(x, y) dx dy; \quad \operatorname{Re}(p_1) > 0, \operatorname{Re}(p_2) > 0 \tag{2.2.1}$$

If $M(p_1, p_2)$ is the Double Mellin transform of $f(x, y)$, then

$$M(p_1, p_2) = \int_0^\infty \int_0^\infty x^{p_1 - 1} y^{p_2 - 1} f(x, y) dx dy; \quad p_1 > 0,\ p_2 > 0 \tag{2.2.2}$$

Generalized hypergeometric function is defined as:

$${}_P F_Q\left[(a_P);(b_Q);z\right] = {}_P F_Q\left[\begin{matrix}(a_P)\\(b_Q)\end{matrix};z\right] = \sum_{n=0}^{\infty} \frac{\prod_{j=1}^{P}(a_j)_n}{\prod_{j=1}^{Q}(b_j)_n} \frac{z^n}{n!}, \tag{2.2.3}$$

Where for brevity, (a_p) denotes the array of parameters $a_1,\ldots,a_p$ with similar interpretation for (b_q) etc. .For further details one can refer Rainville [4].

The following formula is required in the proof:

$$\int_0^\infty \int_0^\infty x^{s-1} y^{t-1} H\left[ax^\lambda, by^\mu\right] dx dy = \frac{a^{-s/\lambda} b^{-t/\mu}}{\lambda \mu} \phi\left(-\frac{s}{\lambda}, -\frac{t}{\mu}\right) \theta_2\left(-\frac{s}{\lambda}\right) \theta_3\left(-\frac{t}{\mu}\right) \tag{2.2.4}$$

The following theorem is also required in the proof:

Theorem: If $F(p_1, p_2)$ is the laplace transform and $M(p_1, p_2)$ is the Mellin transform of $f(t_1, t_2)$, then

$$F(p_1,p_2)=\sum_{s_1=0}^{\infty}\sum_{s_2=0}^{\infty}\frac{(-p_1)^{s_1}}{s_1\,!}\frac{(-p_2)^{s_2}}{s_2\,!}M(s_1+1,s_2+1) \tag{2.2.5}$$

Provided $f(t_1,t_2)$ is continuous for all values of t_1 and t_2, the Laplace transform of $|f(t_1,t_2)|$ exists and the series on the right hand side of $F(p_1,p_2)$ converges.

$H[X]$ Represents the H-function of Fox (1961).

The H-function of two variables (Mittal and Gupta (1972), p.172) using the following notation, which is due essentially to Srivastava and Panda ((1976 a), p.266, eq. (1.5) et seq.)is defined and represented as:

$$H[x,y]=H\left[\begin{smallmatrix}x\\y\end{smallmatrix}\right]=H^{0,n_1:m_2,n_2:m_3,n_3}_{p_1,q_1:p_2,q_2:p_3,q_3}\left[\begin{matrix}x\\y\end{matrix}\middle|\begin{matrix}(a_j;\alpha_j,A_j)_{1,p_1}:(c_j,\gamma_j)_{1,p_2},(e_j,E_j)_{1,p_3}\\(b_j;\beta_j,B_j)_{1,q_1}:(d_j,\delta_j)_{1,q_2},(f_j,F_j)_{1,q_3}\end{matrix}\right]$$

$$=-\frac{1}{4\pi^2}\int_{L_1}\int_{L_2}\phi(\xi,\eta)\,\theta_2(\xi)\,\theta_3(\eta)\,x^{\xi}y^{\eta}\,d\xi\,d\eta \tag{2.2.6}$$

Where

$$\phi(\xi,\eta)=\frac{\prod_{j=1}^{n_1}\Gamma(1-a_j+\alpha_j\xi+A_j\eta)}{\prod_{j=n_1+1}^{p_1}\Gamma(a_j-\alpha_j\xi-A_j\eta)\prod_{j=1}^{q_1}\Gamma(1-b_j+\beta_j\xi+B_j\eta)} \tag{2.2.7}$$

$$\theta_2(\xi)=\frac{\prod_{j=1}^{n_2}\Gamma(1-c_j+\gamma_j\xi)\prod_{j=1}^{m_2}\Gamma(d_j-\delta_j\xi)}{\prod_{j=n_2+1}^{p_2}\Gamma(c_j-\gamma_j\xi)\prod_{j=m_2+1}^{q_2}\Gamma(1-d_j+\delta_j\xi)} \tag{2.2.8}$$

$$\theta_3(\eta)=\frac{\prod_{j=1}^{n_3}\Gamma(1-e_j+E_j\eta)\prod_{j=1}^{m_3}\Gamma(f_j-F_j\eta)}{\prod_{j=n_3+1}^{p_3}\Gamma(e_j-E_j\eta)\prod_{j=m_3+1}^{q_3}\Gamma(1-f_j+F_j\eta)} \tag{2.2.9}$$

2. A Double Hankel Transform:

$$\int_0^\infty\int_0^\infty x^{\rho-1}y^{\sigma-1}F_S\left[(g_R);(k_S);dx^u y^v\right]H_{p,q}^{m,n}\left[cx^\lambda y^\delta \middle|{}_1(g_j,G_j)_p \atop {}_1(h_j,H_j)_q\right]H\left[ax^\gamma,by^\eta\right]dxdy$$

$$=\sum_{r=0}^{\infty}a^{-(\rho+2r)/\gamma}b^{-(\sigma+2r)/\eta}(\gamma\eta)^{-1}f(r)$$

$$H_{p+q_1+q_2+q_3,q+p_1+p_2+p_3}^{m+n_2+n_3,n+m_2+m_3}\left[ca^{-\lambda/\gamma}b^{-\delta/\eta}\middle| {{}_1(g_j,G_j)_n,{}_1(1-d_j-\left(\frac{\rho+r}{\gamma}\right)D_j;\frac{\lambda}{\gamma}D_j)_{m_2}} \atop {{}_1(h_j,H_j)_m,{}_1(1-c_j-\left(\frac{\rho+r}{\gamma}\right)C_j;\frac{\lambda}{\gamma}C_j)_{n_2}}\right.$$

$$\left.{{}_1\left(1-f_j-\left(\frac{\sigma+r}{\eta}\right)F_j,\frac{\delta}{\eta}F_j\right)_{p_3},{}_{n+1}(g_j,G_j)_p,{}_{n_2+1}\left(1-d_j-\left(\frac{\rho+r}{\gamma}\right)D_j,\frac{\lambda}{\gamma}D_j\right)_{q_2}\,{}_1\left(1-b_j-\left(\frac{\rho+r}{\gamma}\right)\beta_j-\left(\frac{\sigma+r}{\eta}\right)B_j+\frac{\lambda}{\gamma}\beta_j+\frac{\delta}{\eta}B_j\right)_{q_1}} \atop {{}_1\left(1-e_j-\left(\frac{\sigma+r}{\eta}\right)E_j,\frac{\delta}{\eta}E_j\right)_{p_3},{}_{m+1}(h_j,H_j)_q,{}_{n_2+1}\left(1-c_j-\left(\frac{\rho+r}{\gamma}\right)C_j,\frac{\lambda}{\gamma}C_j\right)_{p_2}\,{}_1\left(1-a_j-\left(\frac{\rho+r}{\gamma}\right)\alpha_j-\left(\frac{\sigma+r}{\eta}\right)A_j+\frac{\lambda}{\gamma}\alpha_j+\frac{\delta}{\eta}A_j\right)_{p_1}}\right] \quad (2.2.10)$$

Provided,

$$\lambda,\delta>0;\eta>0;\left|\arg c\right|<\frac{1}{2}\Delta\pi,\Delta>0$$

Where
$$f(r)=\frac{\prod_{j=1}^{R}(g_j)_r}{\prod_{j=1}^{S}(k_j)_r}\frac{d^r}{r!}$$

$$\Delta=\sum_{j=1}^{m}H_j-\sum_{j=m+1}^{q}H_j+\sum_{j=1}^{n}G_j-\sum_{j=n+1}^{p}G_j$$

$$\mathrm{Re}[(\rho+r+\gamma(d_i/D_i)+\lambda(h_j/H_j)]>0;i=1,\ldots,m_2;j=1,\ldots,m$$

$$\mathrm{Re}[(\sigma+r+\eta(f_i/F_i)+\delta(h_j/H_j)]>0;i=1,\ldots,m_3;j=1,\ldots,m$$

$$\mathrm{Re}\left[\rho+r-\gamma\left(\frac{1-c_i}{C_i}\right)-\lambda\left(\frac{1-g_j}{G_j}\right)\right]<0;i=1,..,n_2;j=1,\ldots,n$$

$$\mathrm{Re}\left[\sigma+r-\eta\left(\frac{1-e_i}{E_i}\right)-\delta\left(\frac{1-g_j}{G_j}\right)\right]<0;i=1,..,n_3;j=1,\ldots,n$$

$R \leq S$ or $R = S+1$ and $|at^{u+v}| < 1$ [none of $k_j\,(j=1,2,\ldots S)$ is a negative integer or zero].

Proof: To prove (2.2.10), we use series representation for the generalized hypergeometric function, substitute the Mellin-Bernes contour integral for $H[cx^{\lambda}y^{\delta}]$ on the left hand side then interchange the order of contour integral and the (x,y)-integrals. Finally we arrive at our result on evaluating the (x,y)-integral by using the result (2.2.3).

3. Application:

If $f(t_1,t_2) = t_1^{\rho-1}t_2^{\sigma-1}F_S\left[(g_R);(k_S);dx^u y^v\right]H_{p,q}^{m,n}\left[ct_1^{\lambda}t_2^{\delta}\left|{}_1(g_j,G_j)_p \atop {}_1(h_j,H_j)_q\right.\right]H\left[at^{\gamma},bt^{\delta}\right]$;

(2.2.5) becomes the double Laplace-Mellin transform of the product of H-functions of one and two variables and take the following form:

$$\int_0^{\infty}\int_0^{\infty} e^{-P_1t_1-P_2t_2}t_1^{\rho-1}t_2^{\sigma-1}F_S\left[(g_R);(k_S);d\,x^u y^v\right]H_{p,q}^{m,n}\left[ct_1^{\lambda}t_2^{\delta}\left|{}_1(g_j,G_j)_p \atop {}_1(h_j,H_j)_q\right.\right]H\left[at_1^{\gamma},bt_2^{\eta}\right]dt_1dt_2 =$$

$$\sum_{s_1=0}^{\infty}\sum_{s_2=0}^{\infty}\sum_{r=0}^{\infty}\frac{(-P_1)^{s_1}(-P_2)^{s_2}}{s_1!s_2!}a^{-(\rho+r+s_1)/\gamma}b^{-(\sigma+r+s_2)/\eta}(\gamma\eta)^{-1}f(r)$$

$$H_{p+q_1+q_2+q_3,\,q+p_1+p_2+p_3}^{m+n_2+n_3,\,n+m_2+m_3}\left[ca^{-\lambda/\gamma}b^{-\delta/\eta}\left|{}_1(g_j,G_j)_n,{}_1\left(1-d_j-\left(\frac{\rho+r+s_1}{\gamma}\right)D_j;\frac{\lambda}{\gamma}D_j\right)_{m_2} \atop {}_1(h_j,H_j)_m,{}_1\left(1-c_j-\left(\frac{\rho+2r+s_1}{\gamma}\right)C_j;\frac{\lambda}{\gamma}C_j\right)_{m_2}\right.\right.$$

$${}_1\left(1-f_j-\left(\frac{\sigma+r+s_2}{\eta}\right)F_j,\frac{\delta}{\eta}F_j\right)_{p_3},{}_{n+1}(g_j,G_j)_p,{}_{m_2+1}\left(1-d_j-\left(\frac{\rho+r+s_1}{\gamma}\right)D_j,\frac{\lambda}{\gamma}D_j\right)_{q_2}$$
$${}_1\left(1-e_j-\left(\frac{\sigma+r+s_2}{\eta}\right)E_j,\frac{\delta}{\eta}E_j\right)_{p_3},{}_{m+1}(h_j,H_j)_q,{}_{m_2+1}\left(1-c_j-\left(\frac{\rho+r+s_1}{\gamma}\right)C_j,\frac{\lambda}{\gamma}C_j\right)_{p_2}$$

$$\left.\left.{}_1\left(1-b_j-\left(\frac{\rho+r+s_1}{\gamma}\right)\beta_j-\left(\frac{\sigma+r+s_2}{\eta}\right)B_j+\frac{\lambda}{\gamma}\beta_j+\frac{\delta}{\eta}B_j\right)_{q_1} \atop {}_1\left(1-a_j-\left(\frac{\rho+r+s_1}{\gamma}\right)\alpha_j-\left(\frac{\sigma+r+s_2}{\eta}\right)A_j+\frac{\lambda}{\gamma}\alpha_j+\frac{\delta}{\eta}A_j\right)_{p_1}\right]\right. \tag{2.2.11}$$

Provided the conditions are same as that of (2.2.11) with $Re(p_1)>0$, $Re(p_2)>0$.

Proof: In (2.2.1) put

$$f(t_1,t_2)=t_1^{\rho-1}t_2^{\sigma-1}F_S\left[(g_R);(k_S);dx^u y^v\right]H_{p,q}^{m,n}\left[ct_1^{\lambda}t_2^{\delta}\middle|{}_1(g_j,G_j)_p \atop {}_1(h_j,H_j)_q\right]H\left[at^{\gamma},bt^{\eta}\right]$$

and use (2.2.11) to get,

$$M(p_1,p_2)=\sum_{r=0}^{\infty}a^{-(1-\rho-r-s_1)/\gamma}b^{-(1-\sigma-r-s_2)/\eta}(\gamma\eta)^{-1}f(r)$$

$$H_{p+q_1+q_2+q_3,q+p_1+p_2+p_3}^{m+n_2+n_3,n+m_2+m_3}\left[ca^{-\lambda/\gamma}b^{-\delta/\zeta}\middle| {}_1(g_j,G_j)_n,{}_1(1-d_j-\left(\frac{\rho+r+s_1-1}{\gamma}\right)D_j;\frac{\lambda}{\gamma}D_j)_{m_2} \atop {}_1(h_j,H_j)_m,{}_1(1-c_j-\left(\frac{\rho+r+s_1-1}{\gamma}\right)C_j;\frac{\lambda}{\gamma}C_j)_{m_2}\right.$$

$$\begin{matrix} {}_1\left(1-f_j-\left(\frac{\sigma+r+s_2-1}{\eta}\right)F_j,\frac{\delta}{\eta}F_j\right)_{p_3}, {}_{n+1}(g_j,G_j)_p, {}_{n_2+1}\left(1-d_j-\left(\frac{\rho+r+s_1 1}{\gamma}\right)D_j,\frac{\lambda}{\gamma}D_j\right)_{q_2} \\ {}_1\left(1-e_j-\left(\frac{\sigma+r+s_2-1}{\eta}\right)E_j,\frac{\delta}{\eta}E_j\right)_{p_3}, {}_{m+1}(h_j,H_j)_q, {}_{n_2+1}\left(1-c_j-\left(\frac{\rho+r+s_1-1}{\gamma}\right)C_j,\frac{\lambda}{\gamma}C_j\right)_{p_2} \end{matrix}$$

$$\left.\begin{matrix} {}_1\left(1-b_j-\left(\frac{\rho+r+s_1-1}{\gamma}\right)\beta_j-\left(\frac{\sigma+r+s_2-1}{\eta}\right)B_j+\frac{\lambda}{\gamma}\beta_j+\frac{\delta}{\eta}B_j\right)_{q_1} \\ {}_1\left(1-a_j-\left(\frac{\rho+r+s_1-1}{\gamma}\right)\alpha_j-\left(\frac{\sigma+r+s_2-1}{\eta}\right)A_j+\frac{\lambda}{\gamma}\alpha_j+\frac{\delta}{\eta}A_j\right)_{p_1} \end{matrix}\right]$$

Hence $F(p_1,p_2)$ = the right hand side of (2.2.11).

4. Summation formula:

$$\int_0^{\infty}\int_0^{\infty}e^{-P_1t_1-P_2t_2}t_1^{\rho-1}t_2^{\sigma-1}F_S\left[(g_R);(k_S);d\,t_1^u t_2^v\right]H_{p,q}^{m,n}\left[ct_1^{\lambda}t_2^{\delta}\middle|{}_1(g_j,G_j)_p \atop {}_1(h_j,H_j)_q\right]H\left[at_1^{\gamma},bt_2^{\eta}\right]dt_1dt_2=$$

$$\sum_{s_1=0}^{\infty}\sum_{s_2=0}^{\infty}\sum_{r=0}^{\infty}\frac{(-P_1)^{s_1}(-P_2)^{s_2}}{s_1!s_2!}a^{-(\rho+r+s_1)/\gamma}b^{-(\sigma+r+s_2)/\eta}(\gamma\eta)^{-1}f(r)$$

$$H_{p+q_1+q_2+q_3,q+p_1+p_2+p_3}^{m+n_2+n_3,n+m_2+m_3}\left[ca^{-\lambda/\gamma}b^{-\delta/\zeta}\middle| {}_1(g_j,G_j)_n,{}_1(1-d_j-\left(\frac{\rho+r+s_1}{\gamma}\right)D_j;\frac{\lambda}{\gamma}D_j)_{m_2} \atop {}_1(h_j,H_j)_m,{}_1(1-c_j-\left(\frac{\rho+2r+s_1}{\gamma}\right)C_j;\frac{\lambda}{\gamma}C_j)_{m_2}\right.$$

$$\left.\begin{matrix} {}_1\left(1-f_j-\left(\frac{\sigma+r+s_2}{\eta}\right)F_j,\frac{\delta}{\eta}F_j\right)_{p_3}, {}_{n+1}(g_j,G_j)_p, {}_{n_2+1}\left(1-d_j-\left(\frac{\rho+r+s_1}{\gamma}\right)D_j,\frac{\lambda}{\gamma}D_j\right)_{q_2} {}_1\left(1-b_j-\left(\frac{\rho+r+s_1}{\gamma}\right)\beta_j-\left(\frac{\sigma+r+s_2}{\eta}\right)B_j+\frac{\lambda}{\gamma}\beta_j+\frac{\delta}{\eta}B_j\right)_{q_1} \\ {}_1\left(1-e_j-\left(\frac{\sigma+r+s_2}{\eta}\right)E_j,\frac{\delta}{\eta}E_j\right)_{p_3}, {}_{m+1}(h_j,H_j)_q, {}_{n_2+1}\left(1-c_j-\left(\frac{\rho+r+s_1}{\gamma}\right)C_j,\frac{\lambda}{\gamma}C_j\right)_{p_2} {}_1\left(1-a_j-\left(\frac{\rho+r+s_1}{\gamma}\right)\alpha_j-\left(\frac{\sigma+r+s_2}{\eta}\right)A_j+\frac{\lambda}{\gamma}\alpha_j+\frac{\delta}{\eta}A_j\right)_{p_1} \end{matrix}\right]$$

(2.2.12)

Evaluating the left hand side of (2.2.12),using [Srivastava, Gupta and Goyal(1982), eq.(8.5.6),p.150], the following summation formula is obtained:

$$\sum_{r=0}^{\infty} f(r) a^{-\xi} b^{-\eta} P_1^{-\rho-ur-\gamma\xi} P_2^{-\sigma-vr-\eta\zeta}$$

$$H^{m+n_2+n_3,n+2+m_2+m_3}_{p+2+q_1+q_2+q_3,q+p_1+p_2+p_3}\left[cP_1^{-\delta}P_2^{-\eta}ab \left| \begin{matrix} {}_1(g_j,G_j)_n,(1-u-\gamma,\lambda),(1-v-\eta,\delta), \\ {}_1(h_j,H_j)_m,{}_1(1-c_j;\gamma C_j)_{m_2} \end{matrix} \right.\right.$$

$$\left. \begin{matrix} {}_1(1-d_j;\delta D_j)_{m_2},{}_1\left(1-f_j,\eta F_j\right)_{q_3},{}_{n+1}(g_j,G_j)_p,{}_{m_2+1}\left(1-d_j,\delta D_j\right)_{q_2}\ {}_1\left(1-b_j+\xi\beta_j+\eta B_j\right)_{q_1} \\ {}_1\left(1-e_j,\eta E_j\right)_{p_3},{}_{m+1}(h_j,H_j)_q,{}_{m_2+1}\left(1-c_j,\gamma C_j\right)_{p_2}\ {}_1\left(1-a_j+\xi\alpha_j+\eta A_j\right)_{p_1} \end{matrix} \right] =$$

$$\sum_{s_1=0}^{\infty}\sum_{s_2=0}^{\infty}\sum_{r=0}^{\infty} \frac{(-P_1)^{s_1}(-P_2)^{s_2}}{s_1!s_2!} a^{-(\rho+r+s_1)/\gamma} b^{-(\sigma+r+s_2)/\eta} (\gamma\eta)^{-1} f(r)$$

$$H^{m+n_2+n_3,n+m_2+m_3}_{p+q_1+q_2+q_3,q+p_1+p_2+p_3}\left[ca^{-\lambda/\gamma}b^{-\delta/\eta} \left| \begin{matrix} {}_1(g_j,G_j)_n,{}_1(1-d_j-\left(\frac{\rho+r+s_1}{\gamma}\right)D_j;\frac{\lambda}{\gamma}D_j)_{m_2} \\ {}_1(h_j,H_j)_m,{}_1(1-c_j-\left(\frac{\rho+2r+s_1}{\gamma}\right)C_j;\frac{\lambda}{\gamma}C_j)_{m_2} \end{matrix} \right.\right.$$

$$\left. \begin{matrix} {}_1\left(1-f_j-\left(\frac{\sigma+r+s_2}{\eta}\right)F_j,\frac{\delta}{\eta}F_j\right)_{p_3},{}_{n+1}(g_j,G_j)_p,{}_{m_2+1}\left(1-d_j-\left(\frac{\rho+r+s_1}{\gamma}\right)D_j,\frac{\lambda}{\gamma}D_j\right)_{q_2}\ {}_1\left(1-b_j-\left(\frac{\rho+r+s_1}{\gamma}\right)\beta_j-\left(\frac{\sigma+r+s_2}{\eta}\right)B_j+\frac{\lambda}{\gamma}\beta_j+\frac{\delta}{\eta}B_j\right)_{q_1} \\ {}_1\left(1-e_j-\left(\frac{\sigma+r+s_2}{\eta}\right)E_j,\frac{\delta}{\eta}E_j\right)_{p_3},{}_{m+1}(h_j,H_j)_q,{}_{m_2+1}\left(1-c_j-\left(\frac{\rho+r+s_1}{\gamma}\right)C_j,\frac{\lambda}{\gamma}C_j\right)_{p_2}\ {}_1\left(1-a_j-\left(\frac{\rho+r+s_1}{\gamma}\right)\alpha_j-\left(\frac{\sigma+r+s_2}{\eta}\right)A_j+\frac{\lambda}{\gamma}\alpha_j+\frac{\delta}{\eta}A_j\right)_{p_1} \end{matrix} \right]$$

Provided the conditions are same as that of (2.2..11) with $Re(p_1)>0$, $Re(p_2)>0$

5. Special Cases:

(1) In (2.2.11) let $P_1 \to 0$, $P_2 \to 0$ and $\gamma = \eta = 1$, to get the following result:

$$\int_0^{\infty}\int_0^{\infty} t_1^{\rho-1} t_2^{\sigma-1} F_S\left[(g_R);(k_S);d\,x^u y^v\right] H^{m,n}_{p,q}\left[ct_1^{\lambda}t_2^{\delta}\left| \begin{matrix} {}_1(g_j,G_j)_p \\ {}_1(h_j,H_j)_q \end{matrix} \right.\right] H\left[at_1,bt_2\right]dt_1dt_2 =$$

$$\sum_{r=0}^{\infty} a^{-(\rho+r+s_1)} b^{-(\sigma+r+s_2)} f(r)\ H^{m+n_2+n_3,n+m_2+m_3}_{p+q_1+q_2+q_3,q+p_1+p_2+p_3}\left[ca^{-\lambda}b^{-\delta} \left| \begin{matrix} {}_1(g_j,G_j)_n,{}_1(1-d_j-(\rho+r)D_j;\frac{\lambda}{\gamma}D_j)_{m_2} \\ {}_1(h_j,H_j)_m,{}_1(1-c_j-(\rho+r)C_j;\frac{\lambda}{\gamma}C_j)_{m_2} \end{matrix} \right.\right.$$

$$\left.\begin{matrix} {}_1\left(1-f_j-(\sigma+r)F_j,\delta F_j\right)_{p_3},{}_{n+1}(g_j,G_j)_p,{}_{n_2+1}\left(1-d_j-(\rho+r)D_j,\lambda D_j\right)_{q_2} & {}_1\left(1-b_j-(\rho+r)\beta_j-(\sigma+r)B_j+\lambda\beta_j+\delta B_j\right)_{q_1} \\ {}_1\left(1-e_j-(\sigma+r)E_j,\delta E_j\right)_{p_3},{}_{m+1}(h_j,H_j)_q,{}_{n_2+1}\left(1-c_j-(\rho+r)C_j,\lambda C_j\right)_{p_2} & {}_1\left(1-a_j-(\rho+r)\alpha_j-(\sigma+r)A_j+\lambda\alpha_j+\delta A_j\right)_{p_1} \end{matrix}\right] \tag{2.2.13}$$

(2) In (2.2.12) take $p_1 = q_1 = 0$, to get the double Laplace-Hankel transform of the product of three single H-functions of Fox as:

$$\int_0^\infty\int_0^\infty e^{-P_1t_1-P_2t_2}t_1^{\rho-1}t_2^{\sigma-1}F_S\left[(g_R);(k_S);d\,x^u y^v\right]H_{p,q}^{m,n}\left[ct_1^\lambda t_2^\delta \middle| \begin{matrix} {}_1(g_j,G_j)_p \\ {}_1(h_j,H_j)_q \end{matrix}\right]H\left[at_1^\gamma,bt_2^\eta\right]dt_1dt_2 =$$

$$\sum_{s_1=0}^\infty\sum_{s_2=0}^\infty\sum_{r=0}^\infty\frac{(-P_1)^{s_1}(-P_2)^{s_2}}{s_1!s_2!}a^{-(\rho+r+s_1)/\gamma}b^{-(\sigma+r+s_2)/\eta}(\gamma\eta)^{-1}f(r)$$

$$H_{p+q_1+q_2+q_3,q+p_1+p_2+p_3}^{m+n_2+n_3,n+m_2+m_3}\left[ca^{-\lambda/\gamma}b^{-\delta/\eta}\middle| \begin{matrix} {}_1(g_j,G_j)_n,{}_1(1-d_j-\left(\frac{\rho+r+s_1}{\gamma}\right)D_j;\frac{\lambda}{\gamma}D_j)_{m_2} \\ {}_1(h_j,H_j)_m,{}_1(1-c_j-\left(\frac{\rho+2r+s_1}{\gamma}\right)C_j;\frac{\lambda}{\gamma}C_j)_{m_2} \end{matrix}\right.$$

$$\left.\begin{matrix} {}_1\left(1-f_j-\left(\frac{\sigma+r+s_2}{\eta}\right)F_j,\frac{\delta}{\eta}F_j\right)_{p_3},{}_{n+1}(g_j,G_j)_p,{}_{n_2+1}\left(1-d_j-\left(\frac{\rho+r+s_1}{\gamma}\right)D_j,\frac{\lambda}{\gamma}D_j\right)_{q_2} \\ {}_1\left(1-e_j-\left(\frac{\sigma+r+s_2}{\eta}\right)E_j,\frac{\delta}{\eta}E_j\right)_{p_3},{}_{m+1}(h_j,H_j)_q,{}_{n_2+1}\left(1-c_j-\left(\frac{\rho+r+s_1}{\gamma}\right)C_j,\frac{\lambda}{\gamma}C_j\right)_{p_2} \end{matrix}\right] \tag{2.2.14}$$

Provided the conditions are same as that of (2.1) with $p_1 = q_1 = 0$; $Re(p_1)>0$, $Re(p_2)>0$

CHAPTER 3

SECTION I

TWO DIMENSIONAL WEYL FRACTIONAL CALCULUS ASSOCIATED WITH THE WHITTAKER TRANSFORM

This chapter deals with a new theorem concerning the Whittaker transform of two variables. The result is derived by the application of two dimensional Erdelyi-Kober operators of Weyl type. Some known and new special cases are also given in the end.

1. Introduction and Preliminaries

Following Miller ((1975), p.82), let us denoted by A the class of functions f(x, y) which are differentiable any number of times and let their derivatives be

$0(|x|^{-\xi_1},|y|^{-\xi_2})$ for all $\xi_i\,(i=1,2)$ as $x\to\infty, y\to\infty$.

With some modifications the Erdelyi-Kober Operators of Weyl type in two dimensions of a function f are defined as follows:

$$K_x^{\eta,\alpha}K_y^{\delta,\beta}f(x,y)=\frac{(-1)^{m+n}x^{\eta}y^{\delta}}{\Gamma(m+\alpha)\Gamma(n+\beta)}D_{x,y}^{m+n}\int_x^{\infty}\int_y^{\infty}u^{-\eta-\alpha}v^{-\delta-\beta}$$

$$(u-x)^{m+\alpha-1}(v-y)^{n+\beta-1}f(u,v)dudv, \tag{3.1.1}$$

Provided that $f(x,y)\in A;\alpha,\beta$ are real and $m,n=0,1,2,....;$ where

$D_{x,y}^{m+n}$ stands for the operator $\dfrac{\partial^{m+n}}{\partial x^m\partial y^n}$.

For $\alpha>0,\beta>0,m,n=0$, (1.1)becomes a two-dimensional fractional integration operator:

$$K_x^{\eta,\alpha}K_y^{\delta,\beta}f(x,y)=\frac{x^{\eta}y^{\delta}}{\Gamma(m+\alpha)\Gamma(n+\beta)}\int_x^{\infty}\int_y^{\infty}u^{-\eta-\alpha}v^{-\delta-\beta}$$

$$(u-x)^{\alpha-1}(v-y)^{\beta-1}f(u,v)dudv, \tag{3.1.2}$$

If we assume that $\alpha<0,\beta<0$ and m,n are positive integers such that $\alpha+m>0,\beta+n>0$, then (3.1.1) will yield the partial fractional derivatives of $f(x,y)$

The Laplace transform $h(p,q)$ of a function f is defined as in [Ditkin and Prudnikov (1962)].

$$h(p,q) = L[f(x,y); p,q] = \int_0^\infty\int_0^\infty \exp(-px-qy) f(x,y)dxdy \tag{3.1.3}$$

Analogously, the Laplace transform of $f(a\sqrt{x^2-b^2}, c\sqrt{y^2-d^2})$ is defined by the Laplace transform of $F(x,y)$, where

$$F(x,y) = \begin{cases} f(a\sqrt{x^2-b^2}, c\sqrt{y^2-d^2}; x.b>0; y>d>0, \\ 0, \ otherwise. \end{cases} \tag{3.1.4}$$

We now define

$$h_1(p,q) = L[F(x,y); p,q] = \int_b^\infty\int_d^\infty \exp(-px-qy) f(a\sqrt{x^2-b^2}, c\sqrt{y^2-d^2})dxdy \tag{3.1.5}$$

Where $\operatorname{Re}(p) > 0, \operatorname{Re}(q) > 0$.

The Whittaker transform of two variables $g(p,q)$ of a function F is defined by

$$g(p,q) = W_{\lambda_1,\mu_1}^{\lambda,\mu}[F(x,y); \rho,\sigma,p,q] = \int_b^\infty\int_d^\infty (px)^{\rho-1}(qy)^{\sigma-1}$$

$$\exp\left(\frac{1}{2}px + \frac{1}{2}qy\right) W_{\lambda,\mu}(px) W_{\lambda_1,\mu_1}(qy) F(x,y)dxdy \tag{3.1.6}$$

Where $\operatorname{Re}(p) > 0, \operatorname{Re}(q) > 0, |\arg(px)| < \frac{3\pi}{2}, |\arg(qy)| < \frac{3\pi}{2}, g$ exists and belongs to A. Here $W_{\lambda,\mu}(z)$ is the Whittaker's confluent hypergeometric function defined by [Whittaker and Waston(1964), p.340]

$$W_{\lambda,\mu}(z) = \frac{e^{-\frac{1}{2}z} z^k}{\Gamma\left(\frac{1}{2}-\lambda+\mu\right)} \int_0^\infty t^{-\lambda-\frac{1}{2}+\mu} \left(1+\frac{t}{z}\right)^{\lambda+\mu+\frac{1}{2}} e^{-t} dt, \tag{3.1.7}$$

Where $\operatorname{Re}\left(\frac{1}{2}-\lambda+\mu\right) > 0$.

Before presenting the theorem in the next section, we need the generalized Whittaker transform $g_1(p,q)$ of F defined by

$$g_1(p,q) = G^{\lambda,\alpha,\eta,\mu}_{\lambda_1,\beta,\delta,\mu_1}[F(x,y);\rho,\sigma,p,q] = \int_b^\infty\int_d^\infty (px)^{\rho-1}(qy)^{\sigma-1}$$

$$G^{3,1}_{2,3}\left(px\,\Big|^{1+\lambda,\rho}_{\rho-\mu,\frac{1}{2}+\lambda,\frac{1}{2}-\lambda}\right)G^{3,1}_{2,3}\left(qy\,\Big|^{1+\lambda_1,\sigma}_{\sigma-\mu_1,\frac{1}{2}+\lambda_1,\frac{1}{2}-\lambda_1}\right)F(x,y)dxdy \qquad (3.1.8)$$

Where $g_1(p,q)$ exists and belongs to $A, \mathrm{Re}(p)>0, \mathrm{Re}(q)>0$. Here the function $G^{3,1}_{2,3}(z)$ in (3.1.8) is Meijer's G-function.

In general, the G-function is defined by Meijer (1964) by means of the Mellin-Barnes integral

$$G^{m,n}_{p,q}(z) = G^{m,n}_{p,q}\left(z\,\Big|^{a_1,\dots,a_p}_{b_1,\dots,b_q}\right) = \frac{1}{2\pi i}\int_L \chi(s)\,z^s ds\,, \qquad (3.1.9)$$

Where $i=(-1)^{1/2}, z\neq 0$,

$$\chi(s) = \frac{\prod_{j=1}^{m}\Gamma(b_j-s)\prod_{j=1}^{n}\Gamma(1-a_j+s)}{\prod_{j=m+1}^{q}\Gamma(1-b_j+s)\prod_{j=n+1}^{p}\Gamma(a_j-s)}, \qquad (3.1.10)$$

The object of this chapter is to establish a theorem on the Whittaker transform of two variables which extends the result due to Saxena et. al. (1989), Arora et. al. (1985), Raina and Kriyakova (1983).

Theorem: Let

$$g(p,q) = W^{\lambda,\mu}_{\lambda_1,\mu_1}[F(x,y);\rho,\sigma,p,q] = \int_b^\infty\int_d^\infty (px)^{\rho-1}(qy)^{\sigma-1}$$

$$\exp\left(\frac{1}{2}px+\frac{1}{2}qy\right)W_{\lambda,\mu}(px)W_{\lambda_1,\mu_1}(qy)F(x,y)dxdy \qquad (3.1.11)$$

Be the two-dimensional Whittaker transform, for $\alpha>0,\beta>0$, the following result holds:

$$K_p^{\eta,\alpha} K_q^{\delta,\beta}[g(p,q)] = G_{\lambda_1,\beta,\delta,\mu_1}^{\lambda,\alpha,\eta,\mu}[F(x,y);\rho,\sigma,p,q], \qquad (3.1.12)$$

Where R.H.S. of (3.1..12) is defined by (3.1.8)

Proof: Let $\alpha > 0, \beta > 0$. Then in view of (3.1.2) and (3.1.6), we find that

$$K_p^{\eta,\alpha} K_q^{\delta,\beta}[g(p,q)] = \frac{p^\eta q^\delta \Gamma\left(\frac{1}{2}+\mu-\lambda\right)\Gamma\left(\frac{1}{2}-\mu-\lambda\right)\Gamma\left(\frac{1}{2}+\mu_1-\lambda_1\right)\Gamma\left(\frac{1}{2}-\mu_1-\lambda_1\right)}{\Gamma(\alpha)\Gamma(\beta)}$$

$$\int_p^\infty\int_q^\infty u^{-\eta-\alpha} v^{-\delta-\beta}(u-p)^{\alpha-1}(v-q)^{\beta-1} g(u,v)\,dudv =$$

$$\frac{p^\eta q^\delta \Gamma\left(\frac{1}{2}+\mu-\lambda\right)\Gamma\left(\frac{1}{2}-\mu-\lambda\right)\Gamma\left(\frac{1}{2}+\mu_1-\lambda_1\right)\Gamma\left(\frac{1}{2}-\mu_1-\lambda_1\right)}{\Gamma(\alpha)\Gamma(\beta)}\int_p^\infty\int_q^\infty u^{-\eta-\alpha} v^{-\delta-\beta}(u-p)^{\alpha-1}(v-q)^{\beta-1}$$

$$.[\int_b^\infty\int_d^\infty (ux)^{\rho-1}(vy)^{\sigma-1}\exp\left(\frac{1}{2}px+\frac{1}{2}qy\right)W_{\lambda,\mu}(ux)W_{\lambda_1,\mu_1}(vy)F(x,y)\,dxdy]dudv$$

On changing the order of integrations which is permissible, we get

$$K_p^{\eta,\alpha} K_q^{\delta,\beta}[g(p,q)] = \frac{p^\eta q^\delta \Gamma\left(\frac{1}{2}+\mu-\lambda\right)\Gamma\left(\frac{1}{2}-\mu-\lambda\right)\Gamma\left(\frac{1}{2}+\mu_1-\lambda_1\right)\Gamma\left(\frac{1}{2}-\mu_1-\lambda_1\right)}{\Gamma(\alpha)\Gamma(\beta)}$$

$$\int_b^\infty\int_d^\infty x^{\rho-1}y^{\sigma-1}F(x,y)\{\int_p^\infty\int_q^\infty u^{\rho-\eta-\alpha-1}v^{\sigma-\delta-\beta-1}(u-p)^{\alpha-1}(v-q)^{\beta-1}$$

$$\exp\left(\frac{1}{2}ux+\frac{1}{2}vy\right)W_{\lambda,\mu}(ux)W_{\lambda_1,\mu_1}(vy)\,dudv\}dxdy$$

Evaluating the inner integral through the integral [Erdelyi et. Al (1954), p.212, eq. 75]:

$$\int_p^\infty x^{-\rho} e^{\frac{1}{2}ax} W_{\kappa,\lambda}(ax)(x-p)^{\mu-1}dx = \Gamma(\mu)p^{\mu-\rho}$$

$$\Gamma\left(\frac{1}{2}+\lambda-\kappa\right)\Gamma\left(\frac{1}{2}-\lambda-\kappa\right)G_{2,3}^{1,3}\left(ap\,\Big|_{\rho-\mu,\frac{1}{2}+\lambda,\frac{1}{2}-\lambda}^{1+\kappa,\rho}\right),$$

Where

$0<\mathrm{Re}(\mu)<\mathrm{Re}(\rho-\kappa),|\arg(ap)|<\frac{3\pi}{2}$, we obtain the required result.

3. Corollaries:

1. Let

$$g_2(p,q)=W_{\lambda_1,\mu_1}^{\lambda,\mu}[F(x,y);\mu+\frac{1}{2},\mu_1+\frac{1}{2},p,q]=\int_b^\infty\int_d^\infty(px)^{\mu-\frac{1}{2}}(qy)^{\mu_1-\frac{1}{2}}$$

$$\exp\left(\frac{1}{2}px+\frac{1}{2}qy\right)W_{\lambda,\mu}(px)W_{\lambda_1,\mu_1}(qy)F(x,y)dxdy \tag{3.1.13}$$

Exists and belongs to A, then for $\alpha>0,\beta>0$, the following interesting result holds:

$$K_p^{\alpha,-\alpha}K_q^{\beta,-\beta}[g_2(p,q)]=K_p^{\alpha,-\alpha}K_q^{\beta,-\beta}\{W_{\lambda_1,\mu_1}^{\lambda,\mu}[F(x,y);\mu+\frac{1}{2},\mu_1+\frac{1}{2},p,q]=$$

$$W_{\lambda-\frac{\alpha}{2},\mu+\frac{\alpha}{2}}W_{\lambda_1-\frac{\beta}{2},\mu_1+\frac{\beta}{2}}[F(x,y);\mu-\frac{\alpha}{2}+\frac{1}{2},\mu_1-\frac{\beta}{2}+\frac{1}{2},p,q]. \tag{3.1.14}$$

2. Let

$$g_3(p,q)=W_{\lambda_1,\mu_1}^{\lambda,\mu}[F(x,y);\eta+\lambda,\delta+\lambda_1,p,q]=\int_b^\infty\int_d^\infty(px)^{\eta+\lambda-1}(qy)^{\delta+\lambda_1-1}$$

$$\exp\left(\frac{1}{2}px+\frac{1}{2}qy\right)W_{\lambda,\mu}(px)W_{\lambda_1,\mu_1}(qy)F(x,y)dxdy \tag{3.1.15}$$

Exists and belongs to A, then for $\alpha>0,\beta>0$, the following interesting result holds:

$$K_p^{\eta,\alpha}K_q^{\delta,\beta}[g_3(p,q)]=K_p^{\eta,\alpha}K_q^{\delta,\beta}\{W_{\lambda_1,\mu_1}^{\lambda,\mu}[F(x,y);\eta+\lambda,\delta+\lambda_1,p,q]\}=$$

$$W_{\lambda_1-\beta,\mu_1}^{\lambda-\alpha,\mu}[F(x,y);\eta+\lambda,\delta+\lambda_1,p,q]. \tag{3.1.16}$$

4. Special Cases:

If we put $\exp\left(-\frac{1}{2}px-\frac{1}{2}qy\right)$ in place of $\exp\left(\frac{1}{2}px+\frac{1}{2}qy\right)$ in (3.1.11) and use the result:

$$\int_p^\infty x^{-\rho}e^{-\frac{1}{2}ax}W_{\lambda,\mu}(ax)(x-p)^{\sigma-1}dx=\Gamma(\sigma)p^{\sigma-\rho}G_{2,3}^{3,0}\left(ap\left|_{\rho-\sigma,\frac{1}{2}+\mu,\frac{1}{2}-\mu}^{\rho,1-\lambda}\right.\right),$$

We get a result due to Saxena and Ram (1990):

$$K_p^{\eta,\alpha}K_q^{\delta,\beta}[g(p,q)]=G_{\lambda_1,\beta,\delta,\mu_1}^{\lambda,\alpha,\eta,\mu}[F(x,y);\rho,\sigma,p,q]=\int_b^\infty\int_d^\infty(px)^{\rho-1}(qy)^{\sigma-1}$$

$$G_{2,3}^{3,0}\left(px\left|_{1+\eta-\rho,\frac{1}{2}+\mu,\frac{1}{2}-\mu}^{1-\lambda,1+\alpha+\eta-\rho}\right.\right)G_{2,3}^{3,0}\left(qy\left|_{1+\delta-\sigma,\frac{1}{2}+\mu_1,\frac{1}{2}-\mu_1}^{1-\lambda_1,1+\beta+\delta-\sigma}\right.\right)F(x,y)dxdy, \qquad (3.1.17)$$

Further, if we take $\rho=\sigma=1$ in (3.1.17) and use the identity:

$$W_{m+\frac{1}{2},\pm m}(x)=x^{m+\frac{1}{2}}e^{-\frac{1}{2}},$$

The two-dimensional Whittaker transform reduces to a two-dimensional Laplace transform and consiquently, we have a result given by Saxena et. al. (1989). Further if we take $\eta=-\alpha,\delta=-\beta$, we obtain the result due to Arora et. al. (1985). which itself is a generalization of the result given by Raina and Kiriyakova (1983) to which it reduces for $a=c=1,b=d=0$.

SECTION 2

KOBER FRACTIONAL q-INTEGRAL OPERATOR OF THE BASIC ANALOGUE OF THE $\overline{H}$- FUNCTION

In the present chapter an expansion formulae for a basic analogue $\overline{H}$-function have been derived by the applications of the q-Leibniz rule for the type q-derivatives of a product of two functions.

Expansion formulae involving a basic analogue of Fox's H-function, Meijer's G-function and MacRobert's E-function have been derived as special cases of the main results.

1. Introduction

Yadav and Purohit (2008) introduced a new q-extension of the lebniz rule for the derivatives of a product of two basic functions in terms of a finite q-series involving Weyl type q-derivatives of the functions in the following manner:

$$ {}_zD^{\alpha}_{\infty,q}\{U(z)V(z)\}=\sum_{r=0}^{\alpha}\frac{(-1)^r q^{r(r+1)/2}\left(q^{-\alpha};q\right)_r}{(q;q)_r}\,{}_zD^{\alpha-r}_{\infty,q}\{U(z)\}\,{}_zD^{\alpha}_{\infty,q}\left\{V\left(zq^{\alpha-r}\right)\right\} \quad (3.2.1)$$

Where $U(z)$ and $V(z)$ are two functions and the fractional q-differential operator ${}_zD^{\alpha}_{\infty,q}(.)$ of Weyl type is given by

$$ {}_zD^{\alpha}_{\infty,q}\{f(z)\}=\frac{q^{-\alpha(1+\alpha)/2}}{\Gamma_q(-\alpha)}\int_z^{\infty}(t-z)_{-\alpha-1}f(tq^{1+\alpha})d(t;q), \quad (3.2.2)$$

Where $\mathrm{Re}(\alpha)<0$ and

$$(x-y)_v=x^v\prod_{n=0}^{\infty}\left[\frac{1-\left(y/x\right)q^n}{1-\left(y/x\right)q^{v+n}}\right], \quad (3.2.3)$$

The basic integration cf. Gasper and Rehman (1990), is defined as:

$$\int_z^{\infty}f(t)d(t;q)=z(1-q)\sum_{k=1}^{\infty}q^{-k}f\left(zq^{-k}\right). \quad (3.2.4)$$

In view of the relation (3.2.4), operator (3.2.2) can be expressed as:

$$ {}_zD^{\alpha}_{\infty,q}\left\{f(z)\right\}=\frac{q^{\alpha(1-\alpha)/2}z^{-\alpha}(1-q)}{\Gamma_q(-\alpha)}\sum_{k=0}^{\infty}q^{\alpha k}(1-q^{k+1})_{-\alpha-1}f\left(zq^{\alpha-k}\right), \quad (3.2.5) $$

Where $\mathrm{Re}(\alpha)<0$.

In particular, for $f(z)=z^{-p}$, the equation (3.2.5) yields to

$$ {}_zD^{\alpha}_{\infty,q}\left\{z^{-p}\right\}=\frac{\Gamma_q(p+\alpha)}{\Gamma_q(p)}q^{-\alpha p+\alpha(1-\alpha)/2}z^{-p-\alpha}, \quad (3.2.6) $$

Where $\mathrm{Re}(\alpha)<0$.

We shall make use of the following notations and definitions in the sequel:

For real or complex a and $|q|<1$, the q-shifted factorial is defined as:

$$ \left(a;q\right)_n=\begin{cases}1, & if\ n=0\\ (1-a)(1-aq)...(1-aq^{n-1}), & if\ n\in N\end{cases} \quad (3.2.7) $$

In terms of the q-gamma function, (1.7) can be expressed as

$$ (a;q)_n=\frac{\Gamma_q(a+n)(1-q)^n}{\Gamma_q(a)},n>0 \quad (3.2.8) $$

Where the q-gamma function cf. Gasper and Rahman (1990), is given by

$$ \Gamma_q(a)=\frac{(q;q)_\infty}{\left(q^a;q\right)_\infty(1-q)^{a-1}}, \quad (3.2.9) $$

Where $a\neq 0,-1,-2,...$.

The $\overline{H}$-function occurring in the paper will be defined and represented as follows:

$$ \overline{H}^{M,N}_{P,Q}\left[z\right]=\overline{H}^{M,N}_{P,Q}\left[z\Big|\begin{matrix}(a_j;\alpha_j;A_j)_{1,N},(a_j;\alpha_j)_{N+1,P}\\(b_j,\beta_j)_{1,M},(b_j,\beta_j;B_j)_{M+1,Q}\end{matrix}\right]=\frac{1}{2\pi i}\int_{-i\infty}^{i\infty}\overline{\phi}(\xi)z^{\xi}d\xi \quad (3.2.10) $$

where
$$ \overline{\phi}(\xi)=\frac{\prod_{j=1}^{M}\Gamma(b_j-\beta_j\xi)\prod_{j=1}^{N}\left\{\Gamma(1-a_j+\alpha_j\xi)\right\}^{A_j}}{\prod_{j=M+1}^{Q}\left\{\Gamma(1-b_j+\beta_j\xi)\right\}^{B_j}\prod_{j=N+1}^{P}\Gamma(a_j-\alpha_j\xi)} \quad (3.2.11) $$

Which contains fractional powers of the gamma functions. Here, and throughout the paper $a_j (j = 1,..., p)$ and $b_j (j = 1,..., Q)$ are complex parameters, $\alpha_j \geq 0 (j = 1,..., P), \beta_j \geq 0 (j = 1,..., Q)$ (not all zero simultaneously) and exponents $A_j (j = 1,..., N)$ and $B_j (j = N+1,..., Q)$ can take on non integer values.

The following sufficient condition for the absolute convergence of the defining integral for the $\overline{H}$-function given by equation (3.2.1) have been given by Buschman and Srivastava (1990).

$$\Omega \equiv \sum_{j=1}^{M} |\beta_j| + \sum_{j=1}^{N} |A_j \alpha_j| - \sum_{j=M+1}^{Q} |\beta_j B_j| - \sum_{j=N+1}^{P} |\alpha_j| > 0 \qquad (3.2.12)$$

and $|\arg(z)| < \frac{1}{2} \pi \Omega$ (3.2.13)

The behavior of the $\overline{H}$-function for small values of $|z|$ follows easily from a result recently given by Rathie ((1997),p.306,eq.(6.9)).

The basic analogue of the $\overline{H}$-function Buschman and Srivastava (1990) in terms of Mellin-Barnes type basic contour integral is in the following manner:

$$\overline{H}_{p,q}^{m,n}\left[z;q\left|\begin{matrix}(a_j,\alpha_j;A_j)_{1,n},(a_j,\alpha_j)_{n+1,p}\\(b_j,\beta_j)_{1,m},(b_j,\beta_j;B_j)_{m+1,q}\end{matrix}\right.\right] = \frac{1}{2\pi i}\int_C \frac{\prod_{j=1}^{m} G\left(q^{b_j-\beta_j s}\right)\prod_{j=1}^{n} G\left(q^{1-a_j+\alpha_j A_j s}\right)\pi z^s}{\prod_{j=n+1}^{p_i} G\left(q^{b_j-\beta_j B_j s}\right)\prod_{j=m+1}^{q_i} G\left(q^{1-a_{ji}+\alpha_{ji} s}\right)} ds \qquad (3.2.14)$$

Where

$$G(q^{\alpha}) = \left\{\prod_{n=0}^{\infty}\left(1-q^{\alpha+n}\right)\right\}^{-1} = \frac{1}{\left(q^{\alpha};q\right)_{\infty}} \qquad (3.2.15)$$

And $0 \leq m \leq q_i, 0 \leq n \leq p_i; \alpha_j$ and β_j are all positive integers. The contour C is a line parallel to $\mathrm{Re}(ws) = 0$, with indentations, if necessary, in such a manner that all the poles of $G\left(q^{b_j-\beta_j s}\right), 1 \leq j \leq m$, are to the right, and those of $G\left(q^{1-a_j+\alpha_j A_j s}\right), 1 \leq j \leq n$ to the left of C. The integral converges if

$\text{Re}[A_j s\log(z) - \log\sin\pi s] < 0$ for large values of $|s|$ on the contour C. That is, if $|\{\arg(z) - w_2 w_1^{-1}\log|z|\}| < \pi$ where $|q| < 1, \log q = -w = -(w_1 + iw_2), w, w_1, w_2$ are definite quantities. w_1 and w_2 being real.

For $A_j = 1, B_j = 1$, the $\overline{H}$-function reduces to Fox's H-function and eq. (3.2.14) reduces to the q-analogue of the Fox's H-function due to Saxena et. al. (2005), namely

$$H_{p,q}^{m,n}\left[z;q\left|\begin{matrix}(a,\alpha)\\(b,\beta)\end{matrix}\right.\right] = \frac{1}{2\pi i}\int_C \frac{\prod_{j=1}^{m} G\left(q^{b_j-\beta_j s}\right)\prod_{j=1}^{n} G\left(q^{1-a_j-\alpha_j s}\right)\pi z^s}{\prod_{j=m+1}^{q} G\left(q^{1-b_j-\beta_j s}\right)\prod_{j=n+1}^{p} G\left(q^{a_j+\alpha_j s}\right)\sin\pi s} ds, \tag{3.2.16}$$

Where $0 \le m \le q, 0 \le n \le p$ and $\text{Re}[s\log(z) - \log\sin\pi s] < 0$.

For $\alpha_j = \beta_j = 1, j = 1,...,q$ the definition (3.2.16) reduces to the q-analogue of the Meijer's G-function due to Saxena et. al.(2005), namely

$$H_{p,q}^{m,n}\left[z;q\left|\begin{matrix}(a,1)\\(b,1)\end{matrix}\right.\right] = G_{p,q}^{m,n}\left[z;q\left|\begin{matrix}a_1,...,a_p\\b_1,...,b_q\end{matrix}\right.\right] = \frac{1}{2\pi i}\int_C \frac{\prod_{j=1}^{m} G\left(q^{b_j-s}\right)\prod_{j=1}^{n} G\left(q^{1-a_j-s}\right)\pi z^s}{\prod_{j=m+1}^{q} G\left(q^{1-b_j-s}\right)\prod_{j=n+1}^{p} G\left(q^{a_j+s}\right)\sin\pi s} ds$$

(3.2.17)

Where $0 \le m \le q, 0 \le n \le p$ and $\text{Re}[s\log(z) - \log\sin\pi s] < 0$.

Further, if we set $n = 0$ and $m = q$ in the equation (3.2.17), we get the basic analogue of MacRobert's E-function due to Agarwal (1960), namely

$$G_{p,q}^{m,0}\left[z;q\left|\begin{matrix}a_1,...,a_A\\b_1,...,b_B\end{matrix}\right.\right] = E_q\left[q;b_j:p;a_j:z\right] = \frac{1}{2\pi i}\int_C \frac{\prod_{j=1}^{q} G\left(q^{b_j-s}\right)\pi z^s}{\prod_{j=1}^{p} G\left(q^{a_j-s}\right)G\left(q^{b_j-s}\right)\sin\pi s} ds,$$

(3.2.18)

Where $\mathrm{Re}[s\log(z)-\log\sin\pi s]<0$.

The Fox's H-function and Meijer's G-function have been studied in detail by several mathematicians for their theoretical and applications point of view. These functions have found wide ranging applications in mathematical, physical, biological and statistical sciences. It would be interesting to observe that almost all the classical special functions are the particular cases of the Fox's H-function. A detailed account of various classical special functions expressible in terms of Meijer's G-function or Fox's H-function along with their applications to the aforementioned field can be found in the research monographs by Mathai and Saxena (1973,1978).

A new generalization was considered by Saxena et. al. in the form of the q-extansions of the Fox's H-function and Meijer's G-function by means of the Mellin-Bernes type of basic integral. The advantage of these new extansions of the Fox's H-function and Meijer's G-functions lies in the fact that a number of q-special functions including the basic hypergeometric functions, happens to be the particular cases of the $H_q(.)$ and $G_q(.)$ functions, thus widening the scope for further applications. In a paper, (Saxena and Kumar [1990]), besides proving some interesting relations, have established an important limit formula for the $H_q(.)$ functions when q tends to 1. Various basic functions expressible in terms of the basic analogue of Fox's H-function or basic Meijer's G-function with their applications can be found in the research papers due to [Saxena et. al. (2005)] and [Yadav and Purohit (2006)].

In the present chapter, we shall explore the possibility for derivation of some expansion formulae involving the basic analogue of the Fox's H-function by the applications of the q-Leibniz rule for the Weyl type q-derivatives of a product of two functions. We also investigate the expansion formulae involving the basic analogues of Meijer's G-function and MacRobert's E-function.

2. Main Results

In this section, the author will establish certain results associated with the basic analogue of $\overline{H}$-function by assigning suitable values to the function $U(z), V(z)$ and α in the q-Leibniz rule (3.2.1). The main results to be established are as under:

$$\overline{H}^{m+1,n}_{p+1,q+1}\left[\rho\left(zq^{\mu}\right)^{k};q\left|{}^{A^*,(\lambda,k)}_{(\mu+\lambda,k),B^*}\right.\right]=\sum_{R=0}^{\mu}\frac{(-1)^{R}q^{R(R+1)/2+\lambda R}\left(q^{-\mu};q\right)_{R}\left(q^{\lambda};q\right)_{\mu-R}}{(q;q)_{R}}$$

$$\overline{H}^{m+1,n}_{p+1,q+1}\left[\rho\left(zq^{\mu}\right)^{k};q\left|{}^{A^*,(0,k)}_{(R,k),B^*}\right.\right], \tag{3.2.19}$$

Where $0\le m\le q_i, 0\le n\le p_i, \mathrm{Re}[A_j s\log(z)-\log\sin\pi s]<0, k\ge 0$ and ρ being any complex quantity.

$$\overline{H}^{m,n+1}_{p+1,q+1}\left[\rho\left(zq^{\mu}\right)^{k};q\left|{}^{(1-\mu-\lambda,-k;1),A^*}_{B^*,(1-\lambda,-k;1)}\right.\right]=\sum_{R=0}^{\mu}\frac{(-1)^{R}q^{R(R+1)/2+\lambda R}\left(q^{-\mu};q\right)_{R}\left(q^{\lambda};q\right)_{\mu-R}}{(q;q)_{R}}$$

$$\overline{H}^{m,n+1}_{p+1,q+1}\left[\rho\left(zq^{\mu}\right)^{k};q\left|{}^{(1-R,-k),A^*}_{B^*,(1,-k)}\right.\right] \tag{3.2.20}$$

Where $0\le m\le q_i, 0\le n\le p_i, \mathrm{Re}[A_j s\log(z)-\log\sin\pi s]<0, k<0$ and ρ being any complex quantity.

Proof of the main result:

To prove the result (3.2.19) and (3.2.20), we begin with $U(z)=z^{-\lambda}$ and

$$V(z)=\overline{H}^{m,n}_{p,q}\left[\rho z^{k};q\left|{}^{A^*}_{B^*}\right.\right]$$

In equation (3.2.1.) to obtain

$${}_zD^{\mu}_{\infty,q}\left\{z^{-\lambda}I^{m,n}_{p,q}\left[\rho z^{k};q\left|{}^{A^*}_{B^*}\right.\right]\right\}=\sum_{r=0}^{\mu}\frac{(-1)^{R}q^{R(R+1)/2}\left(q^{-\mu};q\right)_{R}}{(q;q)_{R}}\,{}_zD^{\mu-R}_{\infty,q}\left\{z^{-\lambda}\right\}$$

$$ {}_zD^{\alpha}_{\infty,q}\left\{I^{m,n}_{p,q}\left[\rho(z^{\mu-R})^k;q\left|{}^{A^*}_{B^*}\right.\right]\right\} \tag{3.2.21}$$

n view of the definition (3.2.10), the left hand side of equation (3.2.21) becomes

$$ {}_zD^{\mu}_{\infty,q}\left\{z^{-\lambda}I^{m,n}_{p,q}\left[\rho z^k;q\left|{}^{A^*}_{B^*}\right.\right]\right\}$$

$$=\frac{1}{2\pi i}\int_C \frac{\prod_{j=1}^{m}G\left(q^{b_j-\beta_j s}\right)\prod_{j=1}^{n}G\left(q^{1-a_j+\alpha_j A_j s}\right)\pi\rho^s}{\prod_{j=m+1}^{q}G\left(q^{1-b_j+\beta_j B_j s}\right)\prod_{j=n+1}^{p}G\left(q^{a_j-\alpha_j s}\right)G\left(q^{1-s}\right)\sin\pi s}\,{}_zD^{\mu}_{\infty,q}\left\{z^{-(\lambda-ks)}\right\}ds. \tag{3.2.22}$$

On making use of fractional q-derivative formula (3.2.6) in the above equation (3.2.22), we obtain following interesting transformation for the $\overline{H}_q(.)$ function after certain simplifications:

$$ {}_zD^{\mu}_{\infty,q}\left\{z^{-\lambda}I^{m,n}_{p,q}\left[\rho z^k;q\left|{}^{A^*}_{B^*}\right.\right]\right\}=\frac{z^{-\lambda-\mu}q^{-\mu\lambda+\mu(1-\mu)/2}}{(1-q)^{\mu}}I^{m+1,n}_{p+1,q+1}\left[\rho(zq^{\mu})^k q;\left|{}^{A^*,(\lambda,k)}_{(\mu+\lambda,k),B^*}\right.\right], \tag{3.2.23}$$

Where $k \geq 0$.

Again, if we take $k<0$, we obtain the following fractional q-derivative formula for the $\overline{H}_q(.)$ function, namely

$$ {}_zD^{\mu}_{\infty,q}\left\{z^{-\lambda}I^{m,n}_{p,q}\left[\rho z^k;q\left|{}^{A^*}_{B^*}\right.\right]\right\}=\frac{z^{-\lambda-\mu}q^{-\mu\lambda+\mu(1-\mu)/2}}{(1-q)^{\mu}}I^{m,n+1}_{p+1,q+1}\left[\rho(zq^{\mu})^k;q\left|{}^{(1-\mu-\lambda,-k;1),A^*}_{B^*,(1-\lambda,-k;1)}\right.\right]. \tag{3.2.24}$$

We now substitute and replace μ by R and then z by $zq^{\mu-R}$ respectively, in equation (3.2.23) to obtain the following transformation for the $\overline{H}_q(.)$ function:

$$ {}_zD^{R}_{\infty,q}\left\{I^{m,n}_{p,q}\left[\rho\left(zq^{\mu-R}\right)^k;q\Big|^{A*}_{B*}\right]\right\}=\frac{z^{-R}q^{\frac{R(R+1)}{2}-\mu R}}{(1-q)^R}I^{m+1,n}_{p+1,q+1}\left[\rho(zq^{\mu})^k;q\Big|^{A*,(0,k)}_{(R,k),B*}\right] \tag{3.2.25} $$

Further, in view of the result (3.2.6), one can easily obtain the following relation

$$ {}_zD^{\mu-R}_{\infty,q}\left\{z^{-\lambda}\right\}=\frac{\Gamma_q(\lambda+\mu-R)}{\Gamma_q(\lambda)}q^{(\mu-R)(1-\mu-R-2\lambda)/2}z^{-\lambda-\mu+R}. \tag{3.2.26} $$

On substituting the values of various expressions involved in the equation (3.2.21), from equations (3.2.23), (3.2.25) and (3.2.26), we arrive at the main result (3.2.19).

The proof of the result (3.2.20) follows similarly when $k<0$ and by the usages of the transformation formula (3.2.24) and the relation (3.2.26).

3. Special Cases

In this section, we shall consider some special cases of the main results and deduce certain expansion formulae involving the basic analogue of Fox's H-function , basic analogue of Meijer's G-function and basic analogue of MacRobert's E-function.

If we set $A_j=1, B_j=1$, in the main result (3.2.18), we obtain the following expansion formula involving Fox's H-function, namely

$$ H^{m+1,n}_{p+1,q+1}\left[\rho\left(zq^{\mu}\right)^k;q\Big|^{(a_j,\alpha_j)_{1,p},(\lambda,k)}_{(\mu+\lambda,k),(b_j,\beta_j)_{1,q}}\right]=\sum_{R=0}^{\mu}\frac{(-1)^R q^{\frac{R(R+1)}{2}+\lambda R}\left(q^{-\mu};q\right)_R\left(q^{\lambda};q\right)_{\mu-R}}{(q;q)_R} $$

$$ H^{m+1,n}_{p+1,q+1}\left[\rho(zq^{\mu})^k;q\Big|^{(a_j,\alpha_j)_{1,p},(0,k)}_{(R,k),(b_j,\beta_j)_{1,q}}\right], \tag{3.2.27} $$

Where $0\le m\le q, 0\le n\le p, \operatorname{Re}[s\log(z)-\log\sin\pi s]<0, k\ge 0$ and ρ being any complex quantity.

Similarly, for $A_j = 1, B_j = 1$ and $k = -1$, the main result (3.2.19) reduces to yet another expansion formula associated with the basic analogue of Fox's H-function, namely

$$H_{p+1,q+1}^{m,n+1}\left[\rho(zq^{\mu})^{k};q\left|\begin{matrix}(1-\mu-\lambda,-k),(a_j,\alpha_j)_{1,p}\\(b_i,\beta_i)_{1,q},(1-\lambda,-k)\end{matrix}\right.\right]=\sum_{R=0}^{\mu}\frac{(-1)^R q^{\frac{R(R+1)}{2}+\lambda R}(q^{-\mu};q)_R(q^{\lambda};q)_{\mu-R}}{(q;q)_R}$$

$$H_{p+1,q+1}^{m,n+1}\left[\rho(zq^{\mu})^{k};q\left|\begin{matrix}(1-R,-k),(a_j,\alpha_j)_{1,p}\\(b_i,\beta_i)_{1,q},(1,-k)\end{matrix}\right.\right], \tag{3.2.28}$$

Where $0 \le m \le q, 0 \le n \le p, \mathrm{Re}[s\log z - \log\sin\pi s] < 0, k < 0$ and ρ being any complex quantity.

If we set $\alpha_j = \beta_i = 1, j = 1,...,p; i = 1,...,q$ and $k = 1$, in (3.2.27), we obtain the following expansion formula involving Meijer's $G_q(.)$ function, namely

$$G_{p+1,q+1}^{m+1,n}\left[\rho zq^{\mu};q\left|\begin{matrix}(a_j,1)_{1,p},(\lambda,1)\\(\mu+\lambda,1),(b_i,1)_{1,q}\end{matrix}\right.\right]=\sum_{R=0}^{\mu}\frac{(-1)^R q^{\frac{R(R+1)}{2}+\lambda R}(q^{-\mu};q)_R(q^{\lambda};q)_{\mu-R}}{(q;q)_R}$$

$$G_{p+1,q+1}^{m+1,n}\left[\rho zq^{\mu};q\left|\begin{matrix}(a_j,1)_{1,p},(0,1)\\(R,1),(b_i,1)_{1,q}\end{matrix}\right.\right], \tag{3.2.29}$$

Where $0 \le m \le q, 0 \le n \le p, \mathrm{Re}[s\log z - \log\sin\pi s] < 0$ and ρ being any complex quantity.

Similarly, for $\alpha_j = \beta_i = 1, j = 1,...,p; i = 1,...,q$ and $k = -1$, the result (3.2.28) reduces to yet another expansion formula associated with the basic analogue of Meijer's $G_q(.)$ function, namely

$$G_{p+1,q+1}^{m,n+1}\left[\rho/(zq^{\mu});q\left|\begin{matrix}(1-\mu-\lambda,1),(a_j,1)\\(b_i,1)_{1,q},(1-\lambda,1)\end{matrix}\right.\right]=\sum_{R=0}^{\mu}\frac{(-1)^R q^{\frac{R(R+1)}{2}+\lambda R}(q^{-\mu};q)_R(q^{\lambda};q)_{\mu-R}}{(q;q)_R}$$

$$G_{p+1,q+1}^{m,n+1}\left[\rho/(zq^{\mu});q\left|\begin{matrix}(1-R,1),(a_j,1)_{1,p}\\(b_i,1)_{1,q},(1,1)\end{matrix}\right.\right], \tag{3.2.30}$$

Where $0 \le m \le q, 0 \le n \le p, \mathrm{Re}[s\log z - \log\sin\pi s] < 0$ and ρ being any complex quantity.

Finally, if we set $n=0$ and $m=q$, the result (3.2.29), yields to an expansion formula involving MacRobert's $E_q(.)$ function, namely

$$E_q[q+1;b_j,\mu+\lambda:p+1;a_j,\lambda:\rho zq^{\mu}]=\sum_{R=0}^{\mu}\frac{(-1)^R q^{\frac{R(R+1)}{2}+\lambda R}(q^{-\mu};q)_R(q^{\lambda};q)_{\mu-R}}{(q;q)_R}$$

$$E_q[q+1;b_j,R:p+1,a_j,0:\rho zq^{\mu}], \qquad (3.2.31)$$

Where $\mathrm{Re}[s\log z-\log\sin\pi s]<0$ and ρ being any complex quantity.

CHAPTER 4

SECTION I

SOME

MULTIPLICATION FORMULAE

FOR

GENERALIZED HYPERGEOMETRIC FUNCTIONS

In the present chapter, we established two new useful theorems for the generalized differential operator $D^{m}_{k,\alpha,x}$. As an application of our main results, we obtain two multiplication formulae for $\overline{H}$-function. Some known and new special cases are also given in the end.

1. Introduction

(a) Fractional Differential Operator

The Riemann-Liouville fractional derivative of a function $f(z)$ of a complex order μ is defined by Oldham and Spanier (1975) in the following manner:

$$ {}_{\alpha}D_{z}^{\mu}\{f(z)\} = \begin{cases} \frac{1}{\Gamma(-\mu)}\int_{\alpha}^{z}(z-t)^{-\mu-1}f(t)dt, \ \mathrm{Re}(\mu)<0 \\ \frac{d^{m}}{dz^{m}}\,{}_{\alpha}D_{z}^{\mu-m}f(z), \ 0\le \mathrm{Re}(\mu)\le m \end{cases} \quad (4.1.1)$$

Where m is a positive integer. Mishra (1975) has generalized the fractional derivative operator defined by (4.1.1) in the following manner:

$$D_{k,\alpha,x} = x^{k+\alpha}\,D_{x}^{\alpha}, \ \alpha \ne \mu$$

Also we have

$$ {}_{\alpha}D_{z}^{\mu}(x^{\mu-1}) = \frac{d^{\alpha}x^{\mu-1}}{dx^{\alpha}} = \frac{\Gamma(\mu)}{\Gamma(\mu-\alpha)}x^{\mu-\alpha-1}, \ \alpha \ne \mu \quad (4.1.2)$$

$$D^{m}_{k,\alpha,x}(x^{\mu}) = \prod_{p=0}^{m-1}\frac{\Gamma(\mu+pk+1)}{\Gamma(\mu+pk+1-\alpha)}x^{\mu+km}, \ \alpha \ne \mu+1 \quad (4.1.3)$$

Where α and k are not necessarily integers.

Srivastava (1972) introduced the general class of polynomials [Srivastava and Singh (1983)] as

$$S_{m}^{n}[x] = \sum_{k=0}^{[n/m]}\frac{(-n)_{mk}}{k!}A_{n,k}x^{k}, \ n = 0,1,2,... \quad (4.1.4)$$

Where m and n are arbitrary integers and the coefficients $A_{n,k}(n,k\geq 0)$ are arbitrary constant real and complex.

The $\overline{H}$-function occurring in the paper will be defined and represented as follows :

$$\overline{H}_{P,Q}^{M,N}\left[z\right]=\overline{H}_{P,Q}^{M,N}\left[z\left|\begin{matrix}(a_j;\alpha_j;A_j)_{1,N},(a_j;\alpha_j)_{N+1,P}\\(b_j,\beta_j)_{1,M},(b_j,\beta_j;B_j)_{M+1,Q}\end{matrix}\right.\right]=\frac{1}{2\pi i}\int_{-i\infty}^{i\infty}\overline{\phi}(\xi)z^{\xi}d\xi \tag{4.1.5}$$

where
$$\overline{\phi}(\xi)=\frac{\prod_{j=1}^{M}\Gamma(b_j-\beta_j\xi)\prod_{j=1}^{N}\left\{\Gamma(1-a_j+\alpha_j\xi)\right\}^{A_j}}{\prod_{j=M+1}^{Q}\left\{\Gamma(1-b_j+\beta_j\xi)\right\}^{B_j}\prod_{j=N+1}^{P}\Gamma(a_j-\alpha_j\xi)} \tag{4.1.6}$$

Which

contains fractional powers of the gamma functions. Here, and throughout the paper $a_j(j=1,...,p)$ and $b_j(j=1,...,Q)$ are complex parameters, $\alpha_j\geq 0(j=1,...,P),\beta_j\geq 0(j=1,...,Q)$ (not all zero simultaneously) and exponents $A_j(j=1,...,N)$ and $B_j(j=N+1,...,Q)$ can take on non integer values.

The following sufficient condition for the absolute convergence of the defining integral for the $\overline{H}$-function given by equation (4.1.5) have been given by Buschman and Srivastava (1990).

$$\Omega\equiv\sum_{j=1}^{M}\left|\beta_j\right|+\sum_{j=1}^{N}\left|A_j\alpha_j\right|-\sum_{j=M+1}^{Q}\left|\beta_jB_j\right|-\sum_{j=N+1}^{P}\left|\alpha_j\right|>0 \tag{4.1.7}$$

and $$\left|\arg(z)\right|<\frac{1}{2}\pi\Omega \tag{4.1.8}$$

We shall use the following notation:

$A^*=\left(a_j,\alpha_j;A_j\right)_{1,N},\left(a_j,\alpha_j\right)_{N+1,P}$ and $B^*=\left(b_j,\beta_j\right)_{1,M},\left(b_j,\beta_j;B_j\right)_{M+1,Q}$

2. Main Theorems:

To establish the generalized multiplication formulae, we first prove the following theorems:

Theorem 1.

$$D^m_{l,\lambda-\mu,t}\left\{t^{\lambda-1}S^M_N[wt^\rho]f(xt)\right\}=\sum_{k=0}^{[N/M]}\frac{(-N)_{Mk}}{k!}A_{n,k}w^k\sum_{n=0}^{\infty}\frac{(-x)^n}{n!}$$

$$\prod_{p=0}^{m-1}\frac{\Gamma(\lambda+\rho k+pl)}{\Gamma(\mu+\rho k+pl)}t^{\lambda+\rho k+ml-1}\,{}_{m+1}F_m\left[{}^{-n,\lambda+\rho k,\lambda+\rho k+l,\ldots,\lambda+\rho k+(m-1)l}_{\mu+\rho k,\mu+\rho k+l,\ldots,\mu+\rho k+(m-1)l}\right]D^n_x\left\{f(x)\right\} \quad (4.1.9)$$

Theorem 2.

$$D^m_{l,\lambda-\mu,t}\left\{t^{\lambda}S^M_N[wt^\rho]f(xt)\right\}=\sum_{k=0}^{[N/M]}\frac{(-N)_{Mk}}{k!}A_{n,k}w^k$$

$$\sum_{n=0}^{\infty}\frac{(-x)^n}{n!}\prod_{p=0}^{m-1}\frac{\Gamma(\lambda+\rho k+pl)\Gamma(1-\mu-\rho k-pl)_n}{\Gamma(\mu+\rho k+pl)\Gamma(1-\lambda-\rho k-pl)_n}$$

$$t^{\lambda+\rho k+ml-1}\,{}_{m+1}F_m\left[{}^{-n,\lambda+\rho k-n,\lambda+\rho k-n+l,\ldots,\lambda+\rho k-n+(m-1)l}_{\mu+\rho k-n,\mu+\rho k-n+l,\ldots,\mu+\rho k-n+(m-1)l}\right]D^n_x\left\{f(x)\right\} \quad (4.1.10)$$

The theorems 1 and 2 are valid under the following (sufficient) conditions:

$|t|<1,\rho>0,\mathrm{Re}(\mu+\rho k-n+pl)>0\,and\,\mathrm{Re}(\lambda+\rho k+pl)>0$ and assuming that the series involved in (4.1.9) and (4.1.10) are absolutely convergent.

Proof of theorem 1.

Let us consider the well-known Taylor's expansion

$$f(xt)=\sum_{n=0}^{\infty}\frac{(t-1)^n}{n!}x^nD^n_x[f(x)] \quad (4.1.11)$$

Which is a particular case of Lagrange's expansion.

Now multiplying both sides of (4.1.11) by $t^{\lambda-1}S^m_n[wt^\rho]$ and apply the operator $D^m_{l,\lambda-\mu,t}$ both sides, we get

$$D^m_{l,\lambda-\mu,t}\left\{t^{\lambda-1}S^m_n[wt^\rho]f(xt)\right\}=\sum_{k=0}^{[n/m]}\frac{(-n)_{mk}}{k!}A_{n,k}x^k$$

$$\sum_{n=0}^{\infty}\sum_{h=0}^{n}\frac{(-1)^n(-n)_h}{n!h!}x^nD^m_{l,\lambda-\mu,t}\left\{t^{\lambda+k+\rho k-1}\right\}D^n_x\left\{f(x)\right\} \quad (4.1.12)$$

Now using (4.1.3) in R.H.S. of (4.1.12), we easily get the desired result after a little simplification.

Similarly, we can prove Theorem 2 by using the following expansion

$$tf(xt) = \sum_{n=0}^{\infty} \frac{1}{n!}\left(1-\frac{1}{t}\right)^{n} D_x^n\left[x^n f(x)\right] \tag{4.1.13}$$

3. Applications

Next, we establish two multiplication formulas by using Theorem 1 and Theorem 2 respectively. Let

$$f(x) = x^{\sigma-1}\overline{H}[xy] \tag{4.1.14}$$

Now using (4.1.2), we have

$$D_x^n\left\{x^{\sigma-1}\overline{H}[xy]\right\} = x^{\sigma-n-1}\overline{H}_{P+1,Q+1}^{M,N+1}\left[xy\left|\begin{matrix}(1-\sigma,1;1),A^*\\ B^*,(1-\sigma+n,1;1)\end{matrix}\right.\right] \tag{4.1.15}$$

Also in view of (4.1.3), (4.1.4) and (4.1.5), we have

$$D_{l,\lambda-\mu,t}^{m}\left\{t^{\lambda+\sigma-2}x^{\sigma-1}S_n^m[wt^{\rho}]\overline{H}[xt]\right\} = t^{\lambda+\sigma+\rho k+ml-2}\sum_{k=0}^{[n/m]}\frac{(-n)_{mk}}{k!}A_{n,k}w^k x^{\sigma-1}$$

$$\overline{H}_{P+m,Q+m}^{M,N+m}\left[xyt\left|\begin{matrix}(2-\lambda-\sigma-\rho k,1;1),(2-\lambda-\sigma-\rho k-l,1;1),\ldots,(2-\lambda-\sigma-\rho k-(m-1)l,1;1),A^*\\ B^*,(2-\mu-\sigma-\rho k,1;1),(2-\mu-\sigma-\rho k-l,1;1),\ldots,(2-\mu-\sigma-\rho k-(m-1)l,1;1)\end{matrix}\right.\right] \tag{4.1.16}$$

Substituting the values of differential operators (4.1.15) and (4.1.16) in Theorem 1 and Theorem 2 , $y \to \frac{y}{x}$, comparing the coefficients of w_R both sides and replacing $\lambda \to \lambda-\rho R, \mu-\rho R and\, \sigma \to 1-\sigma$ we arrive at the following interesting multiplication formula for $\overline{H}$-function.

Multiplication Formula 1.

$$\overline{H}_{P+m,Q+m}^{M,N+m}\left[yt\left|\begin{matrix}(1-\lambda+\sigma,1;1),(1-\lambda+\sigma-l,1;1),\ldots,(1-\lambda+\sigma-(m-1)l,1;1),A^*\\ B^*,(1-\mu-\sigma,1;1),(1-\mu-\sigma-l,1;1),\ldots,(1-\mu-\sigma-(m-1)l,1;1)\end{matrix}\right.\right]$$

$$= t^{\sigma}\sum_{n=0}^{\infty}\frac{(-1)^n}{n!}\prod_{p=0}^{m-1}\frac{\Gamma(\lambda+pl)}{\Gamma(\mu+pl)}\,{}_{m+1}F_m\left[\begin{matrix}-n,\lambda,\lambda+l,\ldots,\lambda+(m-1)l\\ \mu,\mu+l,\ldots,\mu+(m-1)l\end{matrix}\right]\overline{H}_{P+1,Q+1}^{M,N+1}\left[y\left|\begin{matrix}(\sigma,1;1),A^*\\ B^*,(\sigma+n,1;1)\end{matrix}\right.\right] \tag{4.1.17}$$

Where $|t|<1, \mathrm{Re}(\mu+pl)>0, \mathrm{Re}(\lambda+pl)>0\,(p=0,1,2,\ldots,(m-1))$.

Multiplication formula 2.

$$\overline{H}^{M,N+m}_{P+m,Q+m}\left[yt\left|\begin{matrix}(\sigma-\lambda,1;1),(\sigma-\lambda-l,1;1),\ldots,(\sigma-\lambda-(m-1)l,1;1),A^{*}\\ B^{*},(\sigma-\mu,1;1),(\sigma-\mu-l,1;1),\ldots,(\sigma-\mu-(m-1)l,1;1)\end{matrix}\right.\right]$$

$$=t^{\sigma-1}\sum_{n=0}^{\infty}\frac{(-1)^n}{n!}\prod_{p=0}^{m-1}\frac{\Gamma(\lambda+pl)(1-\mu-pl)_n}{\Gamma(\mu+pl)(1-\lambda-pl)_n}\,{}_{m+1}F_m\left[\begin{matrix}-n,\lambda-n,\lambda-n+l,\ldots,\lambda-n+(m-1)l\\ \mu-n,\mu-n+l,\ldots,\mu-n+(m-1)l\end{matrix}\right]\overline{H}^{M,N+1}_{P+1,Q+1}\left[y\left|\begin{matrix}(\sigma+n,1;1),A^{*}\\ B^{*},(\sigma,1;1)\end{matrix}\right.\right]$$

(4.1.18)

Where $|t|<1, \mathrm{Re}(\mu-n+pl)>0, \mathrm{Re}(\lambda-n+pl)>0\,(p=0,1,2,\ldots,(m-1))$.

4. Special Cases of Multiplication Formula

Since our multiplication formula involve the $\overline{H}$-function, which is general in nature, we can obtain a large number of other multiplication formulae. For the sake of illustration, we briefly mention some of them here.

(i) In (4.1.17), taking $M=1, N=0=P, Q=2, b_2=-\lambda, \beta_1=1, \beta_2=\nu$, the $\overline{H}$-function reduces to generalized Wright-Bessel function $\overline{J}_{\lambda}^{\nu,\mu}$ [Gupta,Jain and Sharma(2003), p.271, eq. (8)] and we get

$$\frac{\Gamma(\lambda-\sigma+\xi)\Gamma(\lambda-\sigma+l+\xi),\ldots,\Gamma(\lambda-\sigma+(m-1)l+\xi)}{\Gamma(\mu+\sigma+\xi),\Gamma(\mu+\sigma+l+\xi),\ldots,\Gamma(\mu+\sigma+(m-1)l+\xi)}\overline{J}_{\lambda}^{\nu,\mu}[z]$$

$$=t^{\sigma}\sum_{n=0}^{\infty}\frac{(-1)^n}{n!}\prod_{p=0}^{m-1}\frac{\Gamma(\lambda+pl)}{\Gamma(\mu+pl)}\,{}_{m+1}F_m\left[\begin{matrix}-n,\lambda,\lambda+l,\ldots,\lambda+(m-1)l\\ \mu,\mu+l,\ldots,\mu+(m-1)l\end{matrix}\right]\overline{H}^{1,1}_{1,3}\left[y\left|\begin{matrix}(\sigma,1;1)\\ (0,1),(-\lambda,\nu,1),(\sigma+n,1,,;1)\end{matrix}\right.\right]$$

(4.1.19)

Where $(1-\nu)>0, (1+\nu)>0, |t|<1, \mathrm{Re}(\mu+pl)>0, \mathrm{Re}(\lambda+pl)>0(p=0,1,2,\ldots m-1)$.

(ii) In (4.1.17), replacing M,N,P,Q by 1,P,P,Q+1 respectively, the $\overline{H}$-function reduces to the Wright's generalized hypergeometric function ${}_P\overline{\psi}_Q$ [Gupta,Jain and Sharma (2003), p.271,eq. (7)]and we get

$$\frac{\Gamma(\lambda-\sigma+\xi)\Gamma(\lambda-\sigma+l+\xi),\ldots,\Gamma(\lambda-\sigma+(m-1)l+\xi)}{\Gamma(\mu+\sigma+\xi),\Gamma(\mu+\sigma+l+\xi),\ldots,\Gamma(\mu+\sigma+(m-1)l+\xi)}\,{}_P\overline{\psi}_Q\left[yt\left|\begin{matrix}(a_j,\alpha_j;A_j)_{1,P}\\ (b_j,\beta_j;B_j)_{1,Q}\end{matrix}\right.\right]$$

$$=t^{\sigma}\sum_{n=0}^{\infty}\frac{(-1)^n}{n!}\prod_{p=0}^{m-1}\frac{\Gamma(\lambda+pl)}{\Gamma(\mu+pl)}\,{}_{m+1}F_m\left[\begin{matrix}-n,\lambda,\lambda+l,...,\lambda+(m-1)l\\ \mu,\mu+l,...,\mu+(m-1)l\end{matrix}\right]\overline{H}^{1,P+1}_{P+1,Q+2}\left[y\left|\begin{matrix}(\sigma,1;1),(a_j,\alpha_j;A_j)_{1,P}\\(0,1),(b_j,\beta_j;B_j)_{1,Q},(\sigma+n,1,;1)\end{matrix}\right.\right]$$

(4.1.20)

Where

$$|t|<1,\mathrm{Re}(\mu+pl)>0,\mathrm{Re}(\lambda+pl)>0(p=0,1,2,...m-1),$$
$$\sum_{j=1}^{P}\alpha_j+1-\sum_{j=1}^{Q}\beta_j\equiv T>0,|\arg z|<\frac{1}{2}T\pi,1+\sum_{j=1}^{Q}\beta_j-\sum_{j=1}^{P}\alpha_j\geq 0.$$

(iii) The function $g_1=(-1)^p g(\gamma,\eta,\tau,p,z)$ [Gupta,Jain and Sharma(2003),p. 271, eq. (10)], where

$$g_1=(-1)^p g(\gamma,\eta,\tau,p,z)=\frac{K_{d-1}\Gamma(p+1)\Gamma\left(\frac{1}{2}+\frac{\tau}{2}\right)}{2^{2+p}\pi^{1/2}\Gamma\left(\gamma-\frac{\tau}{2}\right)\Gamma(\gamma)}\overline{H}^{1,3}_{3,3}\left[-z\left|\begin{matrix}(1-\gamma,1;1),(1-\gamma-\frac{\tau}{2},1;1),(1-\eta,1;1+p)\\(0,1),(-\frac{\tau}{2},1;1),(-\eta,1;1+p)\end{matrix}\right.\right]$$

(4.1.21)

Where $K_d=\dfrac{2^{1-d}\pi^{-d/2}}{\Gamma\left(\frac{d}{2}\right)}$ [Gupta,Jain and Sharma(2003), p.412, eq.(5). (4.1.22)

From this in (4.1.17), we get

$$\frac{K_{d-1}\Gamma(p+1)\Gamma\left(\frac{1}{2}+\frac{\tau}{2}\right)\Gamma(\lambda-\sigma+\xi)\Gamma(\lambda-\sigma-l+\xi),...,\Gamma(\lambda-\sigma-(m-1)l+\xi)}{2^{2+p}\pi^{1/2}\Gamma\left(\gamma-\frac{\tau}{2}\right)\Gamma(\gamma)\Gamma(\mu+\sigma+\xi)\Gamma(\mu+\sigma+l+\xi),...,\Gamma(\mu+\sigma+(m-1)l+\xi)}$$

$$.\overline{H}^{1,3}_{3,3}\left[-yt\left|\begin{matrix}(1-\gamma,1;1),\left(1-\gamma-\frac{\tau}{2},1;1\right),(1-\eta,1;1+p)\\(0,1),\left(-\frac{\tau}{2},1;1\right),(-\eta,1;1+p)\end{matrix}\right.\right]=t^{\sigma}\sum_{n=0}^{\infty}\frac{(-1)^n}{n!}\frac{K_{d-1}\Gamma(p+1)\Gamma\left(\frac{1}{2}+\frac{\tau}{2}\right)}{2^{2+p}\pi^{1/2}\Gamma\left(\gamma-\frac{\tau}{2}\right)\Gamma(\gamma)}$$

$${}_{m+1}F_m\left[\begin{matrix}-n,\lambda,\lambda+l,...,\lambda+(m-1)l\\ \mu,\mu+l,...,\mu+(m-1)l\end{matrix}\right]\overline{H}^{1,4}_{4,4}\left[y\left|\begin{matrix}(1-\gamma,1;1),\left(1-\gamma-\frac{\tau}{2},1;1\right),(1-\eta,1;1+p),(\sigma,1;1)\\(0,1),\left(-\frac{\tau}{2},1;1\right),(-\eta,1,1+p),(\sigma+n,1;1)\end{matrix}\right.\right]\quad(4.1.23)$$

SECTION 2

STUDY OF SOME TRANSFORMATION FORMULAS INVOLVING I-FUNCTION

In the present chapter we establish four transformations formulae of double infinite series involving the I -function. These formulas are then used to obtain double summation formulas for the said function. Our results are quite general in character and a number of summation formulas can be deduced as particular cases. Several interesting special cases of our main finding have been mentioned briefly.

1. Introduction:

The I-function introduced by Saxena(1982) will be represented and defined as follows :

$$I[Z]=I_{p_i,q_i:r}^{m,n}[Z]=I_{p_i,q_i:r}^{m,n}\left[z\left|\begin{matrix}(a_j,\alpha_j)_{1,n},(a_{ji},\alpha_{ji})_{n+1,p_i}\\(b_j,\beta_j)_{1,m},(b_{ji},\beta_{ji})_{m+1,q_i}\end{matrix}\right.\right]=\frac{1}{2\pi\omega}\int_L\chi(\xi)d\xi \tag{4.2.1}$$

where $\omega=\sqrt{-1}$

$$\chi(\xi)=\frac{\prod_{j=1}^{m}\Gamma(b_j-\beta_j\xi)\prod_{j=1}^{n}\Gamma(1-a_j+\alpha_j\xi)}{\sum_{i=1}^{r}\left\{\prod_{j=m+1}^{q_i}\Gamma(1-b_{ji}-\beta_{ji})\prod_{j=n+1}^{p_i}\Gamma(a_{ji},\alpha_{ji})\right\}} \tag{4.2.2}$$

$p_{i,}q_i(i=1,...,r),m,n$ are integers satisfying $0\le n\le p_i$, $0\le m\le q_i,(i=1,...,r),r$ is finite $\alpha_j,\beta_j,\alpha_{ij},\beta_{ji}$ are real and $a_{j,}b_j,a_{ji},b_{ji}$ are complex numbers such that

$\alpha_j(b_h+v)\neq\beta_h(a_j-v-k)$ for $v,k=01,2,...$

We shall use the following notations:

$$A^*=(a_j,\alpha_j)_{1,n},(a_{ji},\alpha_{ji})_{n+1,p_i};B^*=(b_j,\beta_j)_{1,m},(b_{ji},\beta_{ji})_{m+1,q_i}$$

2. Transformation Formulas:

In this section we establish the following four transformation Formulas for the I-function:

First formula

$$\sum_{m,n=0}^{\infty}x^my^nI_{p_i+2,q_i+1:r}^{m,n+2}\left[z\left|\begin{matrix}(1-a-m,\rho),(1-b-n,\sigma),A^*\\B^*,(1-a-b-m-n,\sigma+\rho)\end{matrix}\right.\right]=(x+y-xy)^{-1}$$

$$\left\{x^{s+1}\sum_{s=0}^{\infty}x^{s+1}I_{p_i+2,q_i+1:r}^{m,n+2}\left[z\left|\begin{matrix}(1-a-s,\rho),(1-b,\sigma),A^*\\B^*,(1-a-b-s,\sigma+\rho)\end{matrix}\right.\right]+\sum_{t=0}^{\infty}y^{t+1}I_{p_i+2,q_i+1:r}^{m,n+2}\left[z\left|\begin{matrix}(1-a,\rho),(1-b-t,\sigma),A^*\\B^*,(1-a-b-t,\sigma+\rho)\end{matrix}\right.\right]\right\} \quad (4.2.3)$$

The formula (4.2.3) is valid, if the following (sufficient) conditions are satisfied.

$(i)\ \rho,\sigma>0 \qquad (ii)\ \Omega-\rho-\sigma>0\ ,|\arg z|<\frac{1}{2}(\Omega-\rho-\sigma)\pi$

$$\Omega=\sum_{j=1}^{m}\beta_j+\sum_{j=1}^{n}\alpha_j-\sum_{j=m+1}^{q_i}\beta_{ji}-\sum_{j=n+1}^{p_i}\alpha_{ji}>0 \quad (4.2.4)$$

$(iii)\max\{|x|,|y|\}<1 \quad or\ x=y=1 \quad with \quad Re(a)>1,\mathrm{Re}(b)>1$

Second formula

$$\sum_{m,n=0}^{\infty}\frac{x^m y^n}{m!n!}I_{p_i,q_i:r}^{m,n}\left[z\left|\begin{matrix}(1-a-m-n,u),(1-b-m,v),A^*\\B^*,(1-c-m,\omega)\end{matrix}\right.\right]$$

$$=\sum_{k=0}^{\infty}\frac{1}{k!}(1-y)^{-a}\left(\frac{x}{1-y}\right)^k I_{p_i+2,q_i+1:r}^{m,n+2}\left[z(1-y)^{-u}\left|\begin{matrix}(1-a-k,u),(1-b-k,v),A^*\\B^*,(1-c-k,\omega)\end{matrix}\right.\right] \quad (4.2.5)$$

Provided that

$(i)u,v,\omega>0 \qquad (ii)\ \Omega-\omega>0,|\arg z|<\frac{1}{2}(\Omega-\omega)\pi$

$(iii)\ |x|+|y|<1\ and\ either \left|\frac{x}{1-y}\right|<1\ or\left|\frac{x}{1-y}\right|=1\ with\ \mathrm{Re}(c-a-b)>0$

Third formula

$$\sum_{m,n=0}^{\infty}\frac{x^m y^n}{m!n!}I_{p_i+3,q_i+2:r}^{m,n+3}\left[z\left|\begin{matrix}(1-a-m-n,u),(1-b-m,v),(1-b'-n,\omega)A^*\\B^*,(1-a-m,u),(1-a-n,u)\end{matrix}\right.\right]$$

$$=\sum_{k=0}^{\infty}\frac{1}{k!}(1-x)^{-b}(1-y)^{b'}\left(\frac{xy}{(1-x)(1-y)}\right)^k I_{p_i+2,q_i+1:r}^{m,n+2}\left[z(1-x)^{-v}(1-y)^{-u}\left|\begin{matrix}(1-b-k,v),(1-b'-k,\omega),A^*\\B^*,(1-a-k,u)\end{matrix}\right.\right]$$

(4.2.6)

Provided that

$$(i) u, v, \omega > 0 \qquad (ii)\ \Omega - 2u > 0, |\arg z| < \frac{1}{2}(\Omega - 2u)\pi$$

$$(iii)\ |x| + |y| < 1 \text{ and either } \left|\frac{xy}{(1-x)(1-y)}\right| < 1 \text{ or } \left|\frac{xy}{(1-x)(1-y)}\right| = 1 \text{ with } \operatorname{Re}(a - b - b') > 0$$

Fourth formula

$$\sum_{m,n=0}^{\infty} \frac{x^m y^n}{m!n!} I_{p_i+3,q_i+1:r}^{m,n+3}\left[z \left| \begin{matrix} (1-a-m-n,u),(1-b-m,v),(1-b'-n,\omega)\,A^* \\ B^*,(1-b-b'-m-n,\omega+v) \end{matrix} \right.\right]$$

$$= \sum_{k=0}^{\infty} (1-y)^{-a} \frac{1}{K!}\left(\frac{x-y}{1-y}\right)^k I_{p_i+3,q_i+1:r}^{m,n+3}\left[z(1-y)^{-u} \left| \begin{matrix} (1-a-k,u),(1-b-k,v),(1-b',\omega),A^* \\ B^*,(1-b-b'-k,\omega+v) \end{matrix} \right.\right]$$

(4.2.7)

Provided that

$$(i) u, v, \omega > 0 \qquad (ii)\ \Omega - v - \omega > 0, |\arg z| < \frac{1}{2}(\Omega - v - \omega)\pi$$

$$(iii)\ \max\{|x|,|y|\} < 1 \text{, either } \left|\frac{x-y}{1-y}\right| < 1 \text{ or } \left|\frac{x-y}{1-y}\right| = 1 \text{ with } \operatorname{Re}(b' - a) > 0$$

In all the aforementioned formulas Ω is given by (2.2).

Derivation of the first formula: Using Mellin-Barnes type of contour integral (4.2.1) for the $\overline{H}$-function occurring on the L.H.S. of (4.2.3) and changing the order of integration and summation, we find that L.H.S. of (4.2.3) .

$$= \frac{1}{2\pi i} \int_{-i\infty}^{i\infty} \phi(\xi) z^{\xi} \frac{\Gamma(a+\rho\xi)\Gamma(b+\sigma\xi)}{\Gamma(a+b+(\sigma+\rho)\xi)} F_2\left[a+\rho\xi, b+\sigma\xi, 1, 1,; a+b+(\sigma+\rho)\xi; x, y\right] d\xi$$

(4.2.8)

Now appealing to a known result due to Srivastava [(1972), p.1259, eq. (2.2)]

$$F_2\left[a,b,1,1;a+b;x,y\right] = (x+y-xy)^{-1}\left\{x,\ {}_2F_1\left[a,1;a+b;x\right] + y\ {}_2F_1\left[b,1;a+b;x\right]\right\} \quad (4.2.9)$$

in (4.2.9) , we get L.H.S. of (4.2.3)

$$= \frac{1}{2\pi i} \int_{-i\infty}^{i\infty} \phi(\xi) z^{\xi} \frac{\Gamma(a+\rho\xi)\Gamma(b+\sigma\xi)}{\Gamma(a+b+(\sigma+\rho)\xi)} (x+y-xy)^{-1}$$

$$\{x\,{}_2F_1[a+\rho\xi,1;a+b+(\rho+\sigma)\xi;x] + y\,{}_2F_1[b+\sigma\xi,1;a+b+(\rho+\sigma)\xi;y]\}\,d\xi \quad (4.2.10)$$

Now expressing the ${}_2F_1$ functions in terms of their series and changing the order of integration and summation, and interpreting the result so obtained with the help of (4.2.1), we arrived at the formula (4.2.3).

Derivation of the formulas (4.2.5) to (4.2.7) : The summation formulas (4.2.5), (4.2.6) and (4.2.7) can be developed on the lines similar to the formula (4.2.3) except that, in place of(4.2.9) , here we use the following known results [Erdeyli et.al(1953),p.238,eq.(2),eq.(3) and eq.(1) respectively]:

$$F_2[\alpha,\beta,\beta';\gamma,\beta;x,y] = (1-y)^{-\alpha}\ {}_2F_1\left[\alpha,\beta;\gamma;\frac{x}{(1-y)}\right] \quad (4.2.11)$$

$$F_2[\alpha,\beta,\beta';\alpha,\alpha;x,y] = (1-x)^{-\beta}(1-y)^{-\beta}\ {}_2F_1\left[\beta,\beta';\alpha;\frac{xy}{(1-x)(1-y)}\right] \quad (4.2.12)$$

$$F_2[\alpha,\beta,\beta';\beta+\beta';x,y] = (1-y)^{-\alpha}\ {}_2F_1\left[\alpha,\beta;\beta+\beta';\frac{x-y}{(1-y)}\right] \quad (4.2.13)$$

3. Summation Formulas:

If we take $x = y = 1$ in (4.2.3) and use the well known Gauss's summation theorem, we arrived at the result

$$\sum_{m,n=0}^{\infty} x^m y^n I^{m,n+2}_{p_i+2,q_i+1:r}\left[z\left|\begin{matrix}(1-a-m,\rho),(1-b-n,\sigma),A^* \\ B^*,(1-a-b-m-n,\sigma+\rho)\end{matrix}\right.\right] =$$

$$I^{m,n+2}_{p_i+2,q_i+1:r}\left[z\left|\begin{matrix}(1-a-m,\rho),(2-b,\sigma),A^* \\ B^*,(2-a-b,\sigma+\rho)\end{matrix}\right.\right] + I^{m,n+2}_{p_i+2,q_i+1:r}\left[z\left|\begin{matrix}(2-a,\rho),(1-b,\sigma),A^* \\ B^*,(2-a-b,\sigma+\rho)\end{matrix}\right.\right] \quad (4.2.14)$$

Valid under the conditions of (2.1).

Again if we put $x = y = \frac{1}{2}$ in (2.3), $y = 1-x$ in (4.2.6) and $x = 1$ in (4.2.7) and make use of well known Gauss's summation theorem therein, we shall arrive at the following results:

$$\sum_{m,n=0}^{\infty} \frac{(1/2)^{m+n}}{m!n!} I^{m,n+2}_{p_i+2,q_i+1:r}\left[z \left| \begin{matrix} (1-a-m-n,u),(1-b-m,v),A^* \\ B^*,(1-c-m,\omega) \end{matrix} \right. \right]$$

$$= 2^a I^{m,n+3}_{p_i+3,q_i+2:r}\left[2^u z \left| \begin{matrix} (1-a,u),(1-b,v),(1-c+a+b,\omega-u-v),A^* \\ B^*,(1-c+a,\omega-u),(1-c+b,\omega-v) \end{matrix} \right. \right] \tag{4.2.15}$$

Where $\omega-u-v>0, \omega\neq u \;\; or \;\; \omega\neq v$

And valid under the conditions of (4.2.5)

$$\sum_{m,n=0}^{\infty} \frac{x^m(1-x)^n}{m!n!} I^{m,n+3}_{p_i+3,q_i+2:r}\left[z \left| \begin{matrix} (1-a-m-n,u),(1-b-m,v),(1-b'-n,\omega),A^* \\ B^*,(1-a-m,u)(1-a-n,u) \end{matrix} \right. \right]$$

$$= x^{-b'}(1-x)^{-b} I^{m,n+3}_{p_i+3,q_i+2:r}\left[zx^{-\omega}\left(1-x\right)^{-v} \left| \begin{matrix} (1-b,v),(1-b',\omega),(1-a+b+b',u-v),A^* \\ B^*,(1-a+b,u-v),(1-a-b',u-\omega) \end{matrix} \right. \right] \tag{4.2.16}$$

Where $u-v-\omega>0, \;\; u\neq\omega, \;\; u\neq v$

And valid under the conditions of (4.2.6).

$$\sum_{m,n=0}^{\infty} \frac{y^n}{m!n!} I^{m,n+3}_{p_i+3,q_i+1:r}\left[z \left| \begin{matrix} (1-a-m-n,u),(1-b-m,v),(1-b'-n,\omega),A^* \\ B^*,(1-b-b'-m-n,\omega+v) \end{matrix} \right. \right]$$

$$= (1-y)^{-a} I^{m,n+3}_{p_i+3,q_i+1:r}\left[z(1-y)^{-u} \left| \begin{matrix} (1-a,u),(1-b,v),(1-b'+a,\omega-u),A^* \\ B^*,(1-b-b'+a,\omega-u+v) \end{matrix} \right. \right] \tag{4.2.17}$$

Where $v\neq\omega, v\neq u$ and valid under the conditions (4.2.7).

4. Special Cases:

(i) If we take r=1in (4.2.3),the I-function reduced to Fox's H-function and we get

$$\sum_{m,n=0}^{\infty} x^m y^n H^{m,n+2}_{p_1+2,q_1+1:1}\left[z \left| \begin{matrix} (1-a-m,\rho),(1-b-n,\sigma),(a_{j1},\alpha_{j1})_{1,p_1} \\ (b_{j1},\beta_{j1})_{1,q_1},(1-a-b-m-n,\rho+\sigma) \end{matrix} \right. \right]$$

$$= (x+y-xy)^{-1} \left\{ \sum_{s=0}^{\infty} \begin{matrix} x^{s+1} H^{m,n+2}_{p_1+2,q_1+1:1}\left[z \left| \begin{matrix} (1-a-s,\rho),(1-b-n),(a_{j1},\alpha_{j1})_{1,p_1} \\ (b_{j1},\beta_{j1})_{1,q_1},(1-a-b-s,\rho+\sigma) \end{matrix} \right. \right] \\ +y^{t+1} H^{m,n+2}_{p_1+2,q_1+1:1}\left[z \left| \begin{matrix} (1-a,\rho),(1-b-t,\sigma),(a_{j1},\alpha_{j1})_{1,p_1} \\ (b_{j1},\beta_{j1})_{1,q_1},(1-a-b-t,\rho+\sigma) \end{matrix} \right. \right] \end{matrix} \right\} \tag{4.2.18}$$

Valid under the conditions of (4.2.3).

(ii) If we take $\rho \to 0$ in (4.2.3), we get the following new transformation formula:

$$\sum_{m,n=0}^{\infty} x^m y^n (a)_m I_{p_i+1,q_i+1:r}^{m,n+1}\left[z\left|{}_{B^*,(1-a-b-m-n,\sigma+\rho)}^{(1-b-n,\sigma;1),A^*}\right.\right] = (x+y-xy)^{-1}$$

$$= \left\{\sum_{s=0}^{\infty} x^{s+1}(a)_s I_{p_i+1,q_i+1:r}^{m,n+1}\left[z\left|{}_{B^*,(1-a-b-s,\sigma)}^{(1-b,\sigma),A^*}\right.\right] + \sum_{t=0}^{\infty} y^{t+1} I_{p_i+1,q_i+1:r}^{m,n+1}\left[z\left|{}_{B^*,(1-a-b-t,\sigma)}^{(1-b-t,\sigma),A^*}\right.\right]\right\} \quad (4.2.19)$$

Valid under the conditions of (4.2.3) .

(iii) If we take $\sigma \to 0$ in (4.2.3), we get

$$\sum_{m,n=0}^{\infty} x^m y^n (b)_n I_{p_i+1,q_i+1:r}^{m,n+1}\left[z\left|{}_{B^*,(1-a-b-m-n,\rho)}^{(1-a-s),A^*}\right.\right] =$$

$$(x+y-xy)^{-1}\left\{\sum_{m,n=0}^{\infty} x^{s+1} I_{p_i+1,q_i+1:r}^{m,n+1}\left[z\left|{}_{B^*,(1-a-b-s,\rho)}^{(1-a-s),A^*}\right.\right] + \sum_{m,n=0}^{\infty} y^{t+1}(b)_t I_{p_i+1,q_i+1:r}^{m,n+1}\left[z\left|{}_{B^*,(1-a-b-t,\rho)}^{(1-a),A^*}\right.\right]\right\}$$

(4.2.20)

CHAPTER 5

SECTION 1

STUDY OF UNIFIED THEOREMS INVOLVING THE LAPLACE TRANSFORM WITH APPLICATION

In this chapter, we establish four interesting theorems exhibiting interconnections between images and originals of related functions in the Laplace transform. We also derive six corollaries of the theorems. Further, we obtain five new and general integrals by the application of the theorems. Two known results are also given as a direct consequence of the third theorem. The importance of our findings lies in the fact that they involve the I-function which are very general in nature and are capable of yielding a large number of simpler and useful integrals merely by specializing the parameters in them. These results may find applications in solving certain problems of applied mathematics.

1. Introduction

The Laplace transform occurring in the paper will be defined in the following usual manner:

$$\overline{f}(s) = L\{f(x);s\} = \int_0^\infty e^{-sx} f(x)dx \tag{5.1.1}$$

Where Re(s)>0 and the function f(x) is such that the integral on the R.H.S. of (5.1.1) is absolutely convergent.

The well known Parseval Goldstein theorem for the transform will be in the sequel:

If $\overline{f}_1(s) = L\{f_1(x);s\}$ and $\overline{f}_1(s) = L\{f_1(x);s\}$

Then $\int_0^\infty f_1(x)\overline{f_2}(x)\,dx = \int_0^\infty f_2(x)\overline{f_1}(x)\,dx$ (5.1.2)

The I-function occurring in the paper is defined and represented as follows [Saxena (1982)]:

$$I[z] = I^{m,n}_{p_i,q_i:r}[z] = I^{m,n}_{p_i,q_i:r}\left[z \left| \begin{matrix} (a_j,\alpha_j)_{1,n},(a_{ji},\alpha_{ji})_{n+1,p_i} \\ (b_j,\beta_j)_{1,m},(b_{ji},\beta_{ji})_{m+1,q_i} \end{matrix} \right. \right] = \frac{1}{2\pi\omega}\int_L \phi(\xi)\, z^{\xi}\, d\xi \tag{5.1.3}$$

Where $\omega = \sqrt{-1}$

$$\phi(\xi)=\frac{\prod_{j=1}^{m}\Gamma(b_j-\beta_j\xi)\prod_{j=1}^{n}\Gamma(1-a_j-\alpha_j\xi)}{\sum_{r=1}^{n}\left\{\prod_{j=m+1}^{q_i}\Gamma(1-b_{ji}-\beta_{ji}\xi)\prod_{j=n+1}^{p_i}\Gamma(a_{ji}-\alpha_{ji}\xi)\right\}} \tag{5.1.4}$$

$p_i, q_i\,(i=1,2,...,r), m, n$ are integers satisfying $0\le m\le q_i, 0\le n\le p_i$ r is finite, $\alpha_j,\beta_j,\alpha_{ji},\beta_{ji}$ are real and a_j,b_j,a_{ji},b_{ji} are complex numbers such that

$\alpha_j(b_h+\nu)\ne\beta_h(a_j-\nu-k),\ for\,k=0,1,2,...$

And $\Omega\cong\sum_{j=1}^{m}\beta_j+\sum_{j=1}^{n}\alpha_j-\sum_{j=m+1}^{q_i}\beta_{ji}-\sum_{j=n+1}^{p_i}\alpha_{ji}>0$ (5.1.5)

$$|\arg z|<\frac{1}{2}\pi\Omega \tag{5.1.6}$$

It is assumed that the conditions corresponding appropriately to conditions (5.1.5) and (5.1.6) are satisfied by I-function occurring throughout the paper.

The following Laplace transforms will be required to prove our theorems. They can be computed directly from the defining integral (5.1.3) of the I-function.

$$s^{-\rho}I_{p_i,q_i:r}^{m,n}\left[zs^{-\lambda}\left|\begin{matrix}(a_j,\alpha_j)_{1,n},(a_{ji},\alpha_{ji})_{n+1,p_i}\\(b_j,\beta_j)_{1,m},(b_{ji},\beta_{ji})_{m+1,q_i}\end{matrix}\right.\right]$$

$$= L\left\{s^{\rho-1}I_{p_i,q_i+1:r}^{m,n}\left[zx^{\lambda}\left|\begin{matrix}(a_j,\alpha_j)_{1,n},(a_{ji},\alpha_{ji})_{n+1,p_i}\\(b_j,\beta_j)_{1,m},(b_{ji},\beta_{ji})_{m+1,q_i},(1-\rho,\lambda)\end{matrix}\right.\right];s\right\} \tag{5.1.7}$$

Where $\min\left\{\min_{1\le j\le m}\operatorname{Re}\left(\rho+\lambda\frac{b_j}{\beta_j}\right),\operatorname{Re}(s),\lambda\right\}>0.$

$$s^{-\rho}I_{p_i,q_i:r}^{m,n}\left[zs^{-\lambda}\left|\begin{matrix}(a_j,\alpha_j)_{1,n},(a_{ji},\alpha_{ji})_{n+1,p_i}\\(b_j,\beta_j)_{1,m},(b_{ji},\beta_{ji})_{m+1,q_i}\end{matrix}\right.\right]$$

$$= L\left\{s^{\rho-1}I_{p_i+1,q_i:r}^{m,n}\left[zx^{\lambda}\left|\begin{matrix}(a_j,\alpha_j)_{1,n},(a_{ji},\alpha_{ji})_{n+1,p_i},(\rho,\lambda)\\(b_j,\beta_j)_{1,m},(b_{ji},\beta_{ji})_{m+1,q_i}\end{matrix}\right.\right];s\right\} \tag{5.1.8}$$

Where $\max_{1\le j\le n}\operatorname{Re}\left(\lambda\frac{a_j-1}{\alpha_j}-\rho\right)<0,\{\operatorname{Re}(s),\lambda\}>0$

2. The Theorems:

Theorem 2.1:

If $L\{f(x);s\}=\overline{f}(s)$ (5.1.9)

And

$$L\left\{x^{\rho-1}\overline{f}(x)I^{m,n}_{p_i,q_i+1:r}\left[zx^{\lambda}\left|\begin{smallmatrix}(a_j,\alpha_j)_{1,n},(a_{ji},\alpha_{ji})_{n+1,p_i}\\(b_j,\beta_j)_{1,m},(b_{ji},\beta_{ji})_{m+1,q_i},(1-\rho,\lambda)\end{smallmatrix}\right.\right];s\right\}=h(s) \tag{5.1.10}$$

Then

$$\int_0^{\infty}(x+s)^{-\rho}f(x)I^{m,n}_{p_i,q_i:r}\left[zx^{\lambda}\left|\begin{smallmatrix}(a_j,\alpha_j)_{1,n},(a_{ji},\alpha_{ji})_{n+1,p_i}\\(b_j,\beta_j)_{1,m},(b_{ji},\beta_{ji})_{m+1,q_i}\end{smallmatrix}\right.\right]dx=h(s) \tag{5.1.11}$$

Where $\min\limits_{1\le j\le m}\operatorname{Re}\left(\lambda\dfrac{b_j}{\beta_j}+\rho\right)>0,\ \min\{\operatorname{Re}(s),\lambda\}>0$ and the integrals involved in equations (5.1.9), (5.1.10) and (5.1.11) are absolutely convergent.

On reducing the I-function occurring in (5.1.10) and (5.1.11) to the Wright hypergeometric function, we can easily arrive at the following results:

Corollary 2.1:

If $L\{f(x);s\}=\overline{f}(s)$ (5.1.12)

And

$$L\left\{x^{\rho-1}\overline{f}(x)\,{}_p\psi_{q+1}\left(\begin{smallmatrix}(a_j,\alpha_j)_{1,p}\\(b_j,\beta_j)_{1,q},(\rho,\lambda)\end{smallmatrix};zx^{\lambda}\right);s\right\}=h(s) \tag{5.1.13}$$

Then

$$\int_0^{\infty}(x+s)^{-\rho}f(x)\,{}_p\psi_q\left(\begin{smallmatrix}(a_j,\alpha_j)_{1,p}\\(b_j,\beta_j)_{1,q}\end{smallmatrix};z(x+s)^{-\lambda}\right)dx=h(s) \tag{5.1.14}$$

Where $\min\{\operatorname{Re}(s),\lambda\}>0$ and the integrals involved in equations (5.1.9), (5.1.10) and (5.1.11) are absolutely convergent.

Similarly, reducing I-function in (5.1.11) to the Bessel function [Jain(1992), p.271, eq. (8)], we have the following corollary after a little simplification.

Corollary 2.2:

If $L\{f(x);s\} = \overline{f}(s)$ (5.1.15)

And

$$L\left\{x^{\rho-1}\overline{f}(x)\, I_{0,3}^{1,0}\left[zx^{\lambda}\left|_{(0,1),(-\lambda,\nu),(1-\rho,\lambda)}^{-\quad -\quad -}\right.\right];s\right\} = h(s) \tag{5.1.16}$$

Then

$$\int_0^{\infty}(x+s)^{-\rho} f(x) J_{\lambda}^{\nu}\left(z(x+s)^{-\lambda}\right)dx = h(s) \tag{5.1.17}$$

Where $\min\{\mathrm{Re}(s),\lambda\} > 0$ and the integrals involved in equations (5.1.9), (5.1.10) and (5.1.11) are absolutely convergent.

Theorem 2.2:

If $L\{f(x);s\} = \overline{f}(s)$ (5.1.18)

And

$$L\left\{x^{\rho-1}e^{-ax}\,\overline{f}(x) I_{p_i,q_i+1:r}^{m,n}\left[zx^{\lambda}\left|\begin{matrix}(a_j,\alpha_j)_{1,n},(a_{ji},\alpha_{ji})_{n+1,p_i}\\(b_j,\beta_j)_{1,m},(b_{ji},\beta_{ji})_{m+1,q_i},(1-\rho,\lambda)\end{matrix}\right.\right];s\right\} = h(s) \tag{5.1.19}$$

Then

$$\int_0^{\infty}(x+s)^{-\rho} f(x-a) I_{p_i,q_i:r}^{m,n}\left[z(x+s)^{-\lambda}\left|\begin{matrix}(a_j,\alpha_j)_{1,n},(a_{ji},\alpha_{ji})_{n+1,p_i}\\(b_j,\beta_j)_{1,m},(b_{ji},\beta_{ji})_{m+1,q_i}\end{matrix}\right.\right]dx = h(s) \tag{5.1.20}$$

Where $\min\limits_{1\le j\le m}\mathrm{Re}\left(\lambda\dfrac{b_j}{\beta_j}+\rho\right) > 0,\ \min\{\mathrm{Re}(s),\lambda\} > 0, a \ge 0$ and the integrals involved are absolutely convergent.

The above theorem is a generalization of Theorem 2.1 and reduces to it. On taking a=0.

On reducing the I-function occurring in (5.2.20) to $F(z(x+s)^{-\lambda}, p^{'})$, the poly logarithm of order $p^{'}$ [Erdelyi(1953), p.30, 1.11, eq (14)] with a slight correction of a negative sign, we can easily at the following result.

Corollary 2.3:

If $L\{f(x);s\} = \overline{f}(s)$ (5.1.21)

And

$$L\left\{x^{\rho-1}e^{-ax}\overline{f}(x)I^{1,1+p^{'}}_{1+p^{'},2+q^{'}}\left[zx^{\lambda}\left|\begin{matrix}(1,1),(1,1)_{1,p^{'}}\\(1,1),(1,1)_{1,p^{'}},(1-\rho,\lambda)\end{matrix}\right.\right];s\right\} = h(s) \tag{5.1.22}$$

Then

$$\int_0^{\infty}(x+s)^{-\rho}f(x-a)F\left[z(x+s)^{-\lambda}, p^{'}\right]dx = h(s) \tag{5.1.23}$$

Provided that the integrals involved are absolutely convergent.

Theorem 2.3:

If $L\{f(x);s\} = \overline{f}(s)$ (5.1.24)

And

$$L\left\{x^{\rho-1}e^{-ax}\overline{f}(x)I^{m,n}_{p_i+1,q_i:r}\left[zx^{\lambda}\left|\begin{matrix}(a_j,\alpha_j)_{1,n},(a_{ji},\alpha_{ji})_{n+1,p_i},(\rho,\lambda)\\(b_j,\beta_j)_{1,m},(b_{ji},\beta_{ji})_{m+1,q_i}\end{matrix}\right.\right];s\right\} = h(s) \tag{5.1.25}$$

Then

$$\int_0^{\infty}(x+s)^{-\rho}f(x)I^{m,n}_{p_i,q_i:r}\left[z(x+s)^{\lambda}\left|\begin{matrix}(a_j,\alpha_j)_{1,n},(a_{ji},\alpha_{ji})_{n+1,p_i}\\(b_j,\beta_j)_{1,m},(b_{ji},\beta_{ji})_{m+1,q_i}\end{matrix}\right.\right]dx = h(s) \tag{5.1.26}$$

Where $\max\limits_{1\le j\le m}\operatorname{Re}\left(\lambda\frac{a_j-1}{\alpha_j}-\rho\right)<0$, $\min\{\operatorname{Re}(s),\lambda\}>0, a\ge 0$ and the integrals involved are absolutely convergent. Reducing the I-function involved in (5.1.16), to the Riemann Zeta function $\phi[z(x+s)^{\lambda}, p^{'}]$, [Erdelyi (1953), p.27, 1.11 eq. (1)] and a little simplification leads to:

Corollary 2.4:

If $L\{f(x);s\}=\overline{f}(s)$ (5.1.27)

And

$$L\left\{x^{\rho-1}e^{-ax}\,\overline{f}(x)I^{1,1+p'}_{2+p',1+q'}\left[zx^{-\lambda}\left|\begin{matrix}(1,1),(0,1)_{1,p'},(\rho,\lambda)\\(0,1),(-1,1)_{1,p'},(1-\rho,\lambda)\end{matrix}\right.\right];s\right\}=h(s) \quad (5.1.28)$$

Then

$$\int_0^\infty (x+s)^{-\rho}f(x)\phi\left[z(x+s)^{-\lambda},p'\right]dx=h(s) \quad (5.1.29)$$

Where $\mathrm{Re}(s,\lambda p'+\rho)>0,\lambda>0$, and the integrals involved are absolutely convergent.

Similarly reducing the I-function involved in (5.1.25) to ${}_pF_q$, the generalization of the generalized hypergeometric function, we have the following result:

Corollary 2.5:

If $L\{f(x);s\}=\overline{f}(s)$ (5.1.30)

And

$$L\left\{x^{\rho-1}\overline{f}(x)\,{}_pF_q\left(\begin{matrix}(a_j,\alpha_j)_{1,p}\\(b_j,\beta_j)_{1,q}\end{matrix};-zx^{-\lambda}\right);s\right\}=h(s) \quad (5.1.31)$$

Then

$$\frac{\prod_{j=1}^{q}\Gamma(b_j)}{\prod_{j=1}^{p}\Gamma(a_j)}\int_0^\infty (x+s)^{-\rho}f(x)I^{2,p}_{p,q+2}\left[z(x+s)^{\lambda}\left|\begin{matrix}(a_j,\alpha_j)_{1,p}\\(\rho,\lambda),(0,1),(b_j,\beta_j)_{1,q}\end{matrix}\right.\right]dx=h(s) \quad (5.1.32)$$

Where $\min\{\mathrm{Re}(s),\lambda\}>0$ and the integrals involved are absolutely convergent.

Theorem 2.4:

If $L\{f(x);s\} = \overline{f}(s)$ (5.1.33)

And

$$L\left\{x^{-\rho}\,\overline{f}(x) I^{m,n}_{p_i,q_i:r}\left[zx^{\lambda}\left|\begin{matrix}(a_j,\alpha_j)_{1,n},(a_{ji},\alpha_{ji})_{n+1,p_i}\\(b_j,\beta_j)_{1,m},(b_{ji},\beta_{ji})_{m+1,q_i}\end{matrix}\right.\right];s\right\} = h(s) \tag{5.1.34}$$

Then

$$\int_0^{\infty}(x+s)^{\rho-1}\overline{f}(x) I^{m,n}_{p_i,q_i+1:r}\left[zx^{\lambda}\left|\begin{matrix}(a_j,\alpha_j)_{1,n},(a_{ji},\alpha_{ji})_{n+1,p_i}\\(b_j,\beta_j)_{1,m},(b_{ji},\beta_{ji})_{m+1,q_i},(1-\rho,\lambda)\end{matrix}\right.\right]dx = h(s) \tag{5.1.35}$$

Where $\max\limits_{1\le j\le m}\operatorname{Re}\left(\lambda\dfrac{a_j-1}{\alpha_j}-\rho\right)<0$, $\min\{\operatorname{Re}(s),\lambda\}>0, a\ge 0$ and the integrals involved are absolutely convergent.

Reducing the I-function involved in (5.1.34), to the g- function, [Saxena (1982), p.98, eq. (1.3)], and a little simplification yields our next result:

Corollary 2.6:

If $L\{f(x);s\} = \overline{f}(s)$ (5.1.36)

And

$$L\left\{x^{-\rho}\,\overline{f}(x)\, g\left[\gamma,\eta,\tau,p';zx^{-\lambda}\right];s\right\} = h(s) \tag{5.1.37}$$

Then

$$\int_0^{\infty}(x)^{\rho-1}\overline{f}(x+s) I^{1,2+p'}_{2+p',3+p'}\left[-zx^{\lambda}\left|\begin{matrix}(1-\gamma,1),(1-\gamma+\frac{\tau}{2},1),(1-\eta,1)_{1+p'}\\(0,1),(-\frac{\tau}{2},1),(-\eta,1)_{1+p'},(1-\rho,\lambda)\end{matrix}\right.\right]dx = h(s) \tag{5.1.38}$$

Where $\min\{\operatorname{Re}(s),\lambda\}>0$ and the integrals involved are absolutely convergent.

It may be noted that theorems similar to those given above exist in the literature [Jain (1992), p.359, th.1 (a)].

However, we have studied and established these theorems having I-function as the kernel.

Proof: To prove Theorem 2.1, we require the following result which is obtained by applying the first shifting theorem to (5.1.7):

$$(s+\alpha)^{-\rho} I_{p_i,q_i:r}^{m,n}\left[z(s+\alpha)^{-\lambda}\left|\begin{matrix}(a_j,\alpha_j)_{1,n},(a_{ji},\alpha_{ji})_{n+1,p_i}\\(b_j,\beta_j)_{1,m},(b_{ji},\beta_{ji})_{m+1,q_i}\end{matrix}\right.\right]$$

$$= L\left\{e^{-ax}x^{\rho-1} I_{p_i,q_i+1:r}^{m,n}\left[zx^{\lambda}\left|\begin{matrix}(a_j,\alpha_j)_{1,n},(a_{ji},\alpha_{ji})_{n+1,p_i}\\(b_j,\beta_j)_{1,m},(b_{ji},\beta_{ji})_{m+1,q_i},(1-\rho,\lambda)\end{matrix}\right.\right];s\right\} \qquad (5.1.39)$$

Where $\min\limits_{1\le j\le m} \operatorname{Re}\left(\lambda\frac{b_j}{\beta_j}+\rho\right)>0,\ \min\{\operatorname{Re}(s),\lambda\}>0$.

Again applying the Parseval-Goldstein Theorem to the transform pairs in (5.1.9) and (5.1.39), replacing α by s and making use of (5.1.10), we easily arrive result (5.1.11) under the conditions stated.

For the proof of Theorem 2.2, we first apply the second shifting theorem on (5.1.18), and then proceed in a similar manner as in the proof of Theorem 2.1.

For the proof of Theorem 2.3, the first shifting theorem is applied on result (5.1.20) instead of result (5.1.7), the rest of the proof is on similar lines as that of Theorem 2.1.

For the proof of Theorem 2.4, we apply the first shifting theorem to (5.1.33), and get

$$L\left\{e^{-at}f(x);s\right\}=\overline{f}(\alpha+s) \qquad (5.1.40)$$

Then applying the Parseval-Goldstein theorem on the operational pairs in (5.1.40) and (5.1.34), replacing α by s, and making use of (5.1.34), we easily arrive at the required result (5.1.35) under the conditions stated.

3. Integrals:

By specializing f(x), in the above theorem/ corollaries we can obtain new integrals involving I-functions. Thus, in Theorem 2.1, if we take $f(x)=(x^2+2ax)^{\nu-1/2}$,

The following integral follows after a little simplification with the help of [Srivastava, Gupta and Goyal (1982), p.138, eq. (13):

$$\int_0^\infty (x^2+2ax)^{\nu-1/2}(x+s)^{-\rho} I_{p_i,q_i:r}^{m,n}\left[z(x+s)^{-\lambda}\left|\begin{matrix}(a_j,\alpha_j)_{1,n},(a_{ji},\alpha_{ji})_{n+1,p_i}\\(b_j,\beta_j)_{1,m},(b_{ji},\beta_{ji})_{m+1,q_i}\end{matrix}\right.\right]dx$$

$$=\frac{\sqrt{\pi}}{2\sin\nu\pi}\Gamma(\nu+1/2)(2a)^r\left[\frac{1}{(s-a)^{\rho-2\nu}}\sum_{r=0}^{\infty}\frac{(a/2)^{\nu+2r}}{r!\Gamma(-\nu+r+1)(s-a)^{2r}}\right.$$

$$I_{p_i+1,q_i+1:r}^{m,n+1}\left[z(s-a)^{-\lambda}\left|\begin{matrix}(1-\rho+2\nu-2r,\lambda),(a_j,\alpha_j)_{1,n},(a_{ji},\alpha_{ji})_{n+1,p_i}\\(b_j,\beta_j)_{1,m},(b_{ji},\beta_{ji})_{m+1,q_i},(1-\rho,\lambda)\end{matrix}\right.\right]$$

$$-\frac{1}{(s-a)^{\rho}}\sum_{r=0}^{\infty}\frac{(a/2)^{\nu+2r}}{r!\Gamma(-\nu+r+1)(s-a)^{2r}}$$

$$\left.I_{p_i+1,q_i+1:r}^{m,n+1}\left[z(s-a)^{-\lambda}\left|\begin{matrix}(1-\rho+2\nu-2r,\lambda),(a_j,\alpha_j)_{1,n},(a_{ji},\alpha_{ji})_{n+1,p_i}\\(b_j,\beta_j)_{1,m},(b_{ji},\beta_{ji})_{m+1,q_i},(1-\rho,\lambda)\end{matrix}\right.\right]\right] \tag{5.1.41}$$

Provided $\nu>-1/2$ and $|\arg(a)|<\pi$, $\min\left\{\min_{1\le j\le m}\mathrm{Re}\left(\rho+\lambda\frac{b_j}{\beta_j}\right),\mathrm{Re}(s),\lambda\right\}>0.$

If we reduce the I-functions involved in (5.1.41) to I-function, we get the result in a very elegant form, after a little simplification:

$$\int_0^\infty (x^2+2ax)^{\nu-1/2}(x+s)^{-\rho} I_{p_i,q_i:r}^{m,n}\left[z(x+s)^{-\lambda}\left|\begin{matrix}(a_j,\alpha_j)_{1,n},(a_{ji},\alpha_{ji})_{n+1,p_i}\\(b_j,\beta_j)_{1,m},(b_{ji},\beta_{ji})_{m+1,q_i}\end{matrix}\right.\right]dx$$

$$=\frac{\sqrt{\pi}}{2\sin\nu\pi}\frac{\Gamma(\nu+1/2)(2a)^r}{(s-a)^{\rho-r}}\left(\frac{a}{2(s-a)}\right)^{-\nu}$$

$$\left[I_{1,0;p_i,q_i+1:r;0,2}^{0,1;m,n;1,0}[]\left[\begin{matrix}z(s-a)^{-\lambda}\\-\left(\frac{a}{2(s-a)}\right)^2\end{matrix}\left|\begin{matrix}(1-\rho+2\nu-2r,\lambda),(1-\rho+2\nu-2r,\lambda),(a_j,\alpha_j)_{1,n},(a_{ji},\alpha_{ji})_{n+1,p_i}\\(b_j,\beta_j)_{1,m},(b_{ji},\beta_{ji})_{m+1,q_i},(1-\rho,\lambda),(0,1)(-\nu,1)\end{matrix}\right.\right]\right]$$

$$-I^{0,1;m,n;1,0}_{1,0;p_i,q_i+1:r;0,2}\left[\begin{matrix} z(s-a)^{-\lambda} \\ -\left(\frac{a}{2(s-a)}\right)^2 \end{matrix}\left|\begin{matrix}(1-\rho,\lambda),(1-\rho,\lambda),(a_j,\alpha_j)_{1,n},(a_{ji},\alpha_{ji})_{n+1,p_i} \\ (b_j,\beta_j)_{1,m},(b_{ji},\beta_{ji})_{m+1,q_i},(1-\rho,\lambda),(0,1)(-\nu,1)\end{matrix}\right.\right]\Bigg] \tag{5.1.42}$$

Again taking $f(x)=x^{\nu}$ in Theorem 2.2 yields after a little simplification:

$$\int_0^{\infty}(x-a)^{\nu}(x+s)^{-\rho} I^{m,n}_{p_i,q_i:r}\left[z(x+s)^{-\lambda}\left|\begin{matrix}(a_j,\alpha_j)_{1,n},(a_{ji},\alpha_{ji})_{n+1,p_i} \\ (b_j,\beta_j)_{1,m},(b_{ji},\beta_{ji})_{m+1,q_i}\end{matrix}\right.\right]dx$$

$$= \frac{\Gamma(\nu+1)}{(s+a)^{\rho-\nu-1}} I^{m,n+1}_{p_i+1,q_i+1:r}\left[z(s+a)^{\lambda}\left|\begin{matrix}(2-\rho+\nu,\lambda),(a_j,\alpha_j)_{1,n},(a_{ji},\alpha_{ji})_{n+1,p_i} \\ (b_j,\beta_j)_{1,m},(b_{ji},\beta_{ji})_{m+1,q_i},(1-\rho,\lambda)\end{matrix}\right.\right] \tag{5.1.43}$$

Provided that $\min\left\{\min_{1\le j\le m}\operatorname{Re}\left(\rho-\nu-1+\lambda\frac{b_j}{\beta_j}\right),\operatorname{Re}(\nu+1,s),\lambda\right\}>0.$

Similarly, if we take $f(x)=(1+a/x)^{k/2}\,p_n^k(1+2x/a)$ where $P_n^k(x)$ is the Legendre function [Gupta ,Koul(1977)p.1009,eqn(8.771(1)], in theorem 2.3, simply using [Gradshteyn and Ryzhik (1963),p.216,eq.(16);p.294,eq.(5)], we have an interesting integral:

$$\int_0^{\infty}(1+a/x)^{k/2}P_n^k(1+a/x)(x+a)^{-\rho} I^{m,n}_{p_i,q_i:r}\left[z(x+s)^{-\lambda}\left|\begin{matrix}(a_j,\alpha_j)_{1,n},(a_{ji},\alpha_{ji})_{n+1,p_i} \\ (b_j,\beta_j)_{1,m},(b_{ji},\beta_{ji})_{m+1,q_i}\end{matrix}\right.\right]dx$$

$$=\frac{a^{n+1}}{s^{\rho+n}}\sum_{r=0}^{\infty}\left(\frac{s-a}{s}\right)^r\frac{(n+1-k)_r}{r!}$$

$$I^{m+2,n}_{p_i+2,q_i+2:r}\left[z(s)^{\lambda}\left|\begin{matrix}(a_j,\alpha_j)_{1,n},(a_{ji},\alpha_{ji})_{n+1,p_i},(\rho-k+r,\lambda),(\rho,\lambda) \\ (\rho+n+r,\lambda),(\rho-n-1,\lambda),(b_j,\beta_j)_{1,m},(b_{ji},\beta_{ji})_{m+1,q_i}\end{matrix}\right.\right] \tag{5.1.44}$$

Provided that

Re(k)<1, $\max_{1\le j\le n}\operatorname{Re}\left(\lambda\frac{a_j-1}{\alpha_j}-\rho+n\right)<0,\ \min\{\operatorname{Re}(s),\lambda\}>0,|\arg(a)|>0$

Next, taking $f(x)=x^{\nu}$ in Theorem 2.4, a little simplification yields the following integral:

$$\int_0^{\infty}(x+a)^{-\nu-1}(x)^{-\rho} I^{m,n}_{p_i,q_i+1:r}\left[z(x)^{\lambda}\left|\begin{matrix}(a_j,\alpha_j)_{1,n},(a_{ji},\alpha_{ji})_{n+1,p_i} \\ (b_j,\beta_j)_{1,m},(b_{ji},\beta_{ji})_{m+1,q_i},(1-\rho,\lambda)\end{matrix}\right.\right]dx$$

$$= \frac{}{\Gamma(\nu+1)(s)^{-\rho+\nu+1}} I^{m+1,n}_{p_i,q_i+1:r}\left[z(s+a)^{\lambda} \left|\begin{matrix} (a_j,\alpha_j)_{1,n},(a_{ji},\alpha_{ji})_{n+1,p_i} \\ (\nu-\rho+1,\lambda),(b_j,\beta_j)_{1,m},(b_{ji},\beta_{ji})_{m+1,q_i} \end{matrix}\right.\right] \tag{5.1.45}$$

$$\max_{1\le j\le n} \operatorname{Re}\left(\lambda \frac{a_j-1}{\alpha_j} + \rho - \nu - 1 \right) < 0,\ \lambda > 0 \quad , \qquad \min_{1\le j\le m} \operatorname{Re}\left(\lambda \frac{b_j}{\beta_j} + \rho, s \right) > 0$$

Also, in Theorem 2.3, if we take $f(x) = x^{\eta-1} I^{m,n}_{p_i,q_i:r}\left[zx^{\lambda} \right]$, and reduce the $I^{m,n}_{p_i+1,q_i:r}$ involved in(5.1.25) to $I^{m,0}_{p_i,q_i;r}$, we get a known result [Gupta, Jain and Sharma (2003), p.34], after a little simplification.

Again, if we take $\lambda = 1, \rho = \beta$ and $I^{m,n}_{p_i,q_i:r}$ occurring in (5.1.26) as

$I^{2,0}_{1,2}\left[z(x+s) \left|\begin{matrix} (\alpha,1) \\ (\gamma,1),(\delta,1) \end{matrix}\right.\right]$,

We shall easily arrive at a result by Jain (1992, p.192) after a little simplification.

SECTION 2

SOLUTION OF A CONVOLUTION INTEGRAL EQUATION WITH KERNEL AS A GENERALIZED HYPERGEOMETRIC FUNCTION AND H-FUNCTION OF TWO VARIABLES

In the present chapter, the author gives the solution of a convolution integral equation whose kernel is a generalized hypergeometric function $_pF_Q[.]$ and the H-function of two variables. Some interesting special cases of main result have also been discussed.

1. Introduction:

Generalized hypergeometric function is defined as:

$$ {}_PF_Q\left[(a_P);(b_Q);z\right] = {}_PF_Q\left[\frac{(a_P)}{(b_Q)};z\right] = \sum_{n=0}^{\infty} \frac{\prod_{j=1}^{P}(a_j)_n}{\prod_{j=1}^{Q}(b_j)_n} \frac{z^n}{n!}, \tag{5.2.1} $$

Where for brevity, (a_P) denotes the array of parameters $a_1,\ldots,a_P$ with similar interpretation for (b_Q) etc. . For further details one can refer Rainville (1963).

The H-function of two variables [Mittal and Gupta (1972), p.172] using the following notation, which is due essentially to Srivastava and Panda [(1976a), p.266, eq. (1.5) et seq.]is defined and represented as:

$$ H[x,y] = H\left[\begin{smallmatrix} x \\ y \end{smallmatrix}\right] = H^{0,n_1:m_2,n_2:m_3,n_3}_{p_1,q_1:p_2,q_2:p_3,q_3}\left[\begin{matrix} x \\ y \end{matrix}\middle| \begin{matrix} (a_j;\alpha_j,A_j)_{1,p_1}:(c_j,\gamma_j)_{1,p_2},(e_j,E_j)_{1,p_3} \\ (b_j;\beta_j,B_j)_{1,q_1}:(d_j,\delta_j)_{1,q_2},(f_j,F_j)_{1,q_3} \end{matrix}\right] $$

$$ = -\frac{1}{4\pi^2}\int_{L_1}\int_{L_2}\phi(\xi,\eta)\,\theta_2(\xi)\,\theta_3(\eta)\,x^{\xi}y^{\eta}\,d\xi\,d\eta \tag{5.2.2} $$

Where

$$ \phi(\xi,\eta) = \frac{\prod_{j=1}^{n_1}\Gamma(1-a_j+\alpha_j\xi+A_j\eta)}{\prod_{j=n_1+1}^{p_1}\Gamma(a_j-\alpha_j\xi-A_j\eta)\prod_{j=1}^{q_1}\Gamma(1-b_j+\beta_j\xi+B_j\eta)} \tag{5.2.3} $$

$$ \theta_2(\xi) = \frac{\prod_{j=1}^{n_2}\Gamma(1-c_j+\gamma_j\xi)\prod_{j=1}^{m_2}\Gamma(d_j-\delta_j\xi)}{\prod_{j=n_2+1}^{p_2}\Gamma(c_j-\gamma_j\xi)\prod_{j=m_2+1}^{q_2}\Gamma(1-d_j+\delta_j\xi)} \tag{5.2.4} $$

$$\theta_3(\eta) = \frac{\prod_{j=1}^{n_3} \Gamma(1-e_j+E_j\eta) \prod_{j=1}^{m_3} \Gamma(f_j-F_j\eta)}{\prod_{j=n_3+1}^{p_3} \Gamma(e_j-E_j\eta) \prod_{j=m_3+1}^{q_3} \Gamma(1-f_j+F_j\eta)} \tag{5.2.5}$$

2. The Convolution Integral Equation:

The solution of the following convolution integral equation has been given:

$$\int_0^x (x-t)^{\sigma-1} {}_PF_Q\left[(g_P);(h_Q);a(x-t)^\eta\right] H^{o,n_1:m_2,n_2:m_3,n_3}_{p_1,q_1:p_2,q_2:p_3,q_3}\left[\begin{matrix}(x-t)\\(x-t)\end{matrix}\middle| \begin{matrix}(a_j;\alpha_j,A_j)_{1,p_1}:(c_j,\gamma_j)_{1,p_2}:(e_j,E_j)_{1,p_3}\\(b_j;\beta_j,B_j)_{1,q_1}:(d_j,\delta_j)_{1,q_2}:(f_j,F_j)_{1,q_3}\end{matrix}\right] f(t)dt = g(x) \tag{5.2.6}$$

Where Re(σ)>0, Re(η)>0 and

(i) $R = \sum_{j=1}^{p_1} \alpha_j + \sum_{j=1}^{p_2} \gamma_j - \sum_{j=1}^{q_1} \beta_j - \sum_{j=1}^{q_2} \delta_j < 0$

(ii) $S = \sum_{j=1}^{p_1} A_j + \sum_{j=1}^{p_2} E_j - \sum_{j=1}^{q_1} B_j - \sum_{j=1}^{q_2} F_j < 0$

(iii) $U = -\sum_{j=n_1+1}^{p_1} \alpha_j - \sum_{j=1}^{q_1} \beta_j + \sum_{j=1}^{m_2} \delta_j - \sum_{j=m_2+1}^{q_2} \delta_j + \sum_{j=1}^{n_2} \gamma_j - \sum_{j=n_2+1}^{p_2} \gamma_j > 0$

(iv) $V = -\sum_{j=n_1+1}^{p_1} A_j - \sum_{j=1}^{q_1} B_j + \sum_{j=1}^{m_3} F_j - \sum_{j=m_3+1}^{q_3} F_j + \sum_{j=1}^{n_3} E_j - \sum_{j=n_3+1}^{p_3} E_j > 0$

(V) $|\arg x| < \frac{1}{2} U\pi$

(vi) $|\arg y| < \frac{1}{2} V\pi$

Solution: In order to solve (5.2.6), we first take Laplace transform of both sides of (5.2.6), we get

$$\int_0^\infty e^{-px}\{\int_0^x (x-t)^{\sigma-1}\,{}_PF_Q\left[(g_R);(h_S);a(x-t)^\eta\right]$$

$$H^{0,n_1:m_2,n_2:m_3,n_3}_{p_1,q_1:p_2,q_2:p_3,q_3}\left[\begin{matrix}(x-t)\\(x-t)\end{matrix}\middle|\begin{matrix}(a_j;\alpha_j,A_j)_{1,p_1}:(c_j,\gamma_j)_{1,p_2}:(e_j,E_j)_{1,p_3}\\(b_j;\beta_j,B_j)_{1,q_1}:(d_j,\delta_j)_{1,q_2}:(f_j,F_j)_{1,q_3}\end{matrix}\right]f(t)\,dt\}\,dx=\int_0^\infty e^{-px}g(x)dx$$

$$f(r)H^{0,n_1:m_2,n_2:m_3,n_3}_{p_1,q_1:p_2,q_2:p_3,q_3}\left[\begin{matrix}1\\1\end{matrix}\middle|\begin{matrix}(a_j;\alpha_j,A_j)_{1,p_1}:(c_j,\gamma_j)_{1,p_2}:(e_j,E_j)_{1,p_3}\\(b_j;\beta_j,B_j)_{1,q_1}:(d_j,\delta_j)_{1,q_2}:(f_j,F_j)_{1,q_3}\end{matrix}\right]\int_0^\infty e^{-px}\{\int_0^x (x-t)^{\sigma+\eta r+\xi+\zeta-1}f(t)\,dt\}\,dx=\overline{g}(p)$$

Changing the order of integration, we get

$$f(r)H^{0,n_1:m_2,n_2:m_3,n_3}_{p_1,q_1:p_2,q_2:p_3,q_3}\left[\begin{matrix}1\\1\end{matrix}\middle|\begin{matrix}(a_j;\alpha_j,A_j)_{1,p_1}:(c_j,\gamma_j)_{1,p_2}:(e_j,E_j)_{1,p_3}\\(b_j;\beta_j,B_j)_{1,q_1}:(d_j,\delta_j)_{1,q_2}:(f_j,F_j)_{1,q_3}\end{matrix}\right]\int_0^\infty f(t)\{\int_t^\infty e^{-px}(x-t)^{\sigma+\eta r+\xi+\zeta-1}\,dx\}\,dt=\overline{g}(p)$$

Putting (x-t)=u, we obtain

$$f(r)H^{0,n_1:m_2,n_2:m_3,n_3}_{p_1,q_1:p_2,q_2:p_3,q_3}\left[\begin{matrix}1\\1\end{matrix}\middle|\begin{matrix}(a_j;\alpha_j,A_j)_{1,p_1}:(c_j,\gamma_j)_{1,p_2}:(e_j,E_j)_{1,p_3}\\(b_j;\beta_j,B_j)_{1,q_1}:(d_j,\delta_j)_{1,q_2}:(f_j,F_j)_{1,q_3}\end{matrix}\right]\int_0^\infty f(t)\{\int_0^\infty e^{-p(u+t)}u^{\sigma+\eta r+\xi+\zeta-1}\,du\}\,dt=\overline{g}(p)$$

$$f(r)H^{0,n_1:m_2,n_2:m_3,n_3}_{p_1,q_1:p_2,q_2:p_3,q_3}\left[\begin{matrix}1\\1\end{matrix}\middle|\begin{matrix}(a_j;\alpha_j,A_j)_{1,p_1}:(c_j,\gamma_j)_{1,p_2}:(e_j,E_j)_{1,p_3}\\(b_j;\beta_j,B_j)_{1,q_1}:(d_j,\delta_j)_{1,q_2}:(f_j,F_j)_{1,q_3}\end{matrix}\right]\int_0^\infty e^{-pt}f(t)\{\int_0^\infty e^{-pu}u^{\sigma+\eta r+\xi+\zeta-1}\,du\}\,dt=\overline{g}(p)$$

$$f(r)H^{0,n_1:m_2,n_2:m_3,n_3}_{p_1,q_1:p_2,q_2:p_3,q_3}\left[\begin{matrix}1\\1\end{matrix}\middle|\begin{matrix}(a_j;\alpha_j,A_j)_{1,p_1}:(c_j,\gamma_j)_{1,p_2}:(e_j,E_j)_{1,p_3}\\(b_j;\beta_j,B_j)_{1,q_1}:(d_j,\delta_j)_{1,q_2}:(f_j,F_j)_{1,q_3}\end{matrix}\right]\overline{f}(p)\frac{\Gamma(\sigma+\eta r+\xi+\zeta)}{p^{\sigma+\eta r+\xi+\zeta}}=\overline{g}(p)$$

$$\frac{f(r)\overline{f}(p)H_2\left[p^{-1},p^{-1}\right]}{p^{\sigma+\eta r}}=\overline{g}(p)$$

Or $$\overline{f}(p)=\frac{p^{\sigma+\eta r}}{f(r)H_2\left[p^{-1},p^{-1}\right]} \qquad (5.2.7)$$

Where $\overline{f}(p)$ and $\overline{g}(p)$ denote the Laplace transform of f(t) and g(t), respectively, and

$$H_2\left[p^{-1},p^{-1}\right]=H^{0,n_1:m_2,n_2:m_3,n_3}_{p_1+1,q_1:p_2,q_2:p_3,q_3}\left[\begin{matrix}p^{-1}\\p^{-1}\end{matrix}\middle|\begin{matrix}(1-\sigma-\eta r,1,1),(a_j;\alpha_j,A_j)_{1,p_1}:(c_j,\gamma_j)_{1,p_2}:(e_j,E_j)_{1,p_3}\\(b_j;\beta_j,B_j)_{1,q_1}:(d_j,\delta_j)_{1,q_2}:(f_j,F_j)_{1,q_3}\end{matrix}\right] \qquad (5.2.8)$$

And $f(r)=\sum_{r=0}^{\infty}\frac{\prod_{j=1}^{p}(g_j)_r}{\prod_{j=1}^{q}(h_j)_r}\frac{a^r}{r!}$. (5.2.9)

A series expansion for $H_2\left[p^{-1},p^{-1}\right]$ can be obtain as a special case of series expansion of H[x,y] [Srivastava, Gupta and Goyal (1982),eq.(6.2.1),p.84]and, since this specialization leads to a power series, the series representation for the reciprocal can be formed. To do this we note that

$$H_2\left[p^{-1},p^{-1}\right]=\sum_{M,N=0}^{\infty}C_{M,N}(-p)^{-M-N}\frac{\Gamma(\sigma+\eta r+M+N)}{M!N!} \quad (5.2.10)$$

Where $C_{M,N}=\phi(M,N)\theta_2(M)\theta_3(N)$ (5.2.11)

Where $\phi(\xi,\zeta),\theta_2(\xi),\theta_3(\zeta)$ are given by (5.2.3), (5.2.4), (5.2.5) respectively.

From the well -known rearrangement property [Rainville (1963), p.56]

$$\sum_{M,N=0}^{\infty}F(M,N)=\sum_{M=0}^{\infty}\sum_{N=0}^{M}f(M-N,N) \quad (5.2.12)$$

We can rewrite H_2 as a single (power) series in the form

$$H_2\left[p^{-1},p^{-1}\right]=\sum_{\nu=0}^{\infty}h_\nu p^{-\nu} \quad (5.2.13)$$

Where $h_\nu=\frac{(-1)^\nu\Gamma(\sigma+\eta r+\nu)}{\nu!}\sum_{\mu=0}^{\nu}\binom{\nu}{\mu}C_{\nu-\mu,\mu}$ (5.2.14)

If k denotes the least value of ν for which $h_\nu\neq 0$, then

$$H_2\left[p^{-1},p^{-1}\right]=p^{-k}\sum_{n=0}^{\infty}h_{k+n}p^{-n} \quad (5.2.15)$$

So that if we let the coefficients H_λ be determined by the relation

$$\left[\sum_{n=0}^{\infty} h_{k+n} p^{-n}\right]^{-1} = \sum_{\lambda=0}^{\infty} H_{\lambda} p^{-\lambda} \tag{5.2.16}$$

Then (2.2) becomes

$$\overline{f}(p) = p^{\sigma+k}\overline{g}(p)\frac{1}{f(r)}\sum_{\lambda=0}^{\infty} H_{\lambda} p^{-\lambda}$$

$$= p^{-(\rho-k-\sigma-\eta r)}\left[\frac{1}{f(r)}\sum_{\lambda=0}^{\infty} H_{\lambda} p^{-\lambda}\right]\left[p^{\rho}\overline{g}(p)\right] \tag{5.2.17}$$

Consequently, on taking the inverse Laplace transform of (5.2.17) and applying its convolution theorem, we obtain the following:

Theorem: If Re(σ)>0, Re(η)>0, $g^{r}(0)=0$ for 0≤r≤ρ,ρ an integer, Re(ρ-k-σ-ηr)>0, then under suitable restrictions on the parameters of the H-functions of two variables occurring in (5.2.6) [obtainable easily from the set of conditions (i) to (vi) mentioned with (5.2.6)]the solution to the convolution integral equation (5.2.6) is given by

$$f(t) = \int_{0}^{t}(t-x)^{\rho-k-\sigma-\eta r-1}V(t-x)g^{\rho}(x)dx \tag{5.2.18}$$

Where $V(x) = \dfrac{1}{f(r)}\displaystyle\sum_{\lambda=0}^{\infty}\frac{H_{\lambda}x^{\lambda}}{\Gamma(\rho-k+\lambda-\sigma-\eta r)}$ (5.2.19)

The coefficients H_{λ} being defined by the recurrences

$H_k H_0=1$, and for μ>0 by $\displaystyle\sum_{\lambda=0}^{\mu} H_{\lambda} h_{\mu+k-\lambda} = 0$ (5.2.20)

And the power series coefficients h_{ν} being given by (5.2.14).

3. Special Cases:

(i) If we take $p_1=q_1=0$ in (5.2.6), the H-function of two variables reduces to the product of two single H-function of Fox as:

$$\int_0^x (x-t)^{\sigma-1} {}_PF_Q\left[(g_P);(h_Q);(x-t)^\eta\right] H^{o,n_2}_{p_2,q_2}\left[(x-t)\left|\begin{smallmatrix}(c_j,\gamma_j)_{1,p_2}\\(d_j,\delta_j)_{1,q_2}\end{smallmatrix}\right.\right]$$

$$H^{o,n_3}_{p_3,q_3}\left[(x-t)\left|\begin{smallmatrix}(e_j,E_j)_{1,p_3}\\(f_j,F_j)_{1,q_3}\end{smallmatrix}\right.\right] f(t)dt = g(x) \quad (5.2.21)$$

Where $g^{(r)}(0)=0$ for $0\le r\le \rho$, ρ an integer and $Re(\sigma)>0$, $Re(\eta)>0$, has its solution given by

$$f(t) = \int_0^t (t-x)^{\rho-k-\sigma-\eta r-1}\omega(t-x) g^{\rho}(x)dx \quad (5.2.22)$$

Where $Re(\rho-k-\sigma-\eta r)>0$ and

$$\omega(x) = \frac{1}{f(r)}\sum_{\lambda=0}^{\infty}\frac{H_\lambda x^\lambda}{\Gamma(\rho-k+\lambda-\sigma-\eta r)} \quad (5.2.23)$$

The H_λ being determined by (5.2.20), h_v being given by (5.2.14), and the coefficients in (5.2.11) reduces to the form:

$$C_{v-\mu,\mu} = \theta_2(v-\mu)\theta_3(\mu)$$

(ii) If we put r=0 in (5.2.6), we get the result due to Buschman, Koul and Gupta (1977) in some slight different form:

$$\int_0^x (x-t)^{\sigma-1} H^{o,n_1:m_2,n_2:m_3,n_3}_{p_1,q_1:p_2,q_2:p_3,q_3}\left[\begin{smallmatrix}(x-t)\\(x-t)\end{smallmatrix}\left|\begin{smallmatrix}(a_j;\alpha_j,A_j)_{1,p_1}:(c_j,\gamma_j)_{1,p_2}:(e_j,E_j)_{1,p_3}\\(b_j;\beta_j,B_j)_{1,q_1}:(d_j,\delta_j)_{1,q_2}:(f_j,F_j)_{1,q_3}\end{smallmatrix}\right.\right] f(t)dt = g(x)$$

Where $g^{(r)}(0)=0$ for $0\le r\le \rho$, ρ an integer and $Re(\sigma)>0$, $Re(\eta)>0$, has its solution given by

$$f(t) = \int_0^t (t-x)^{\rho-k-\sigma-1}\chi(t-x) g^{\rho}(x)dx \quad (5.2.24)$$

Where $Re(\rho-k-\sigma)>0$ and

$$\chi(x) = \frac{1}{f(r)}\sum_{\lambda=0}^{\infty}\frac{H_\lambda x^\lambda}{\Gamma(\rho-k+\lambda-\sigma)} \quad (5.2.25)$$

The H_λ being determined by (5.2.20), h_v being given by (5.2.14).

CHAPTER 6

SECTION 1

COMPOSITION

OF

GENERALIZED FRACTIONAL INTEGRALS

INVOLVING PRODUCT

OF

GENERALIZED HYPERGEOMETRIC FUNCTIONS

We derive three compositions of the fractional integral operators associated with a product of I-function and a general class of polynomials due to Srivastava. The result obtained are of general character and apply for the general composition of the various fractional integration operators associated with special functions of mathematical Physics and Chemistry appeared in various recent publications.

1. Introduction and Preliminaries:

The composition of the following integral operators defined by means of the following integral equations will be derived.

$$Y_{\alpha}^{h}\{f(x)\}=x^{-h-\alpha}A_{\alpha}\{x^{h}f(x)\}=\frac{x^{-h-\alpha}}{\Gamma(\alpha)}\int_{0}^{x}(x-s)^{\alpha-1}I_{p_i,q_i:r}^{m,n}\left[z\left(1-\frac{s}{x}\right)^{c}\right] S_{b}^{a}\left[y\left(1-\frac{s}{x}\right)^{\sigma}\right]s^{h}f(s)\,ds \tag{6.1.1}$$

$$V_{\beta}^{\lambda}\{f(x)\}=x^{-\lambda}B_{\beta}\{x^{-\lambda-\beta}f(x)\}=\frac{x^{-\lambda}}{\Gamma(\beta)}\int_{x}^{\infty}(s-x)^{\beta-1}I_{p_i',q_i':r}^{m',n'}\left[z'\left(1-\frac{x}{s}\right)^{d}\right] S_{f}^{e}\left[y'\left(1-\frac{s}{x}\right)^{\rho}\right]x^{-\lambda-\beta}f(s)\,ds \tag{6.1.2}$$

The I-function appearing in (6.1.1) and (6.1.2) has been introduced by Saxena(1982) and defined as follows:

$$I[Z]=I_{p_i,q_i:r}^{m,n}[Z]=I_{p_i,q_i:r}^{m,n}\left[z\left|\begin{matrix}(a_j,\alpha_j)_{1,n},(a_{ji},\alpha_{ji})_{n+1,p_i}\\(b_j,\beta_j)_{1,m},(b_{ji},\beta_{ji})_{m+1,q_i}\end{matrix}\right.\right]=\frac{1}{2\pi\omega}\int_{L}\chi(s)\,ds \tag{6.1.3}$$

Where $\omega=\sqrt{-1}$

$$\chi(s)=\frac{\prod_{j=1}^{m}\Gamma(b_j-\beta_j s)\prod_{j=1}^{n}\Gamma(1-a_j+\alpha_j s)}{\sum_{i=1}^{r}\left\{\prod_{j=m+1}^{q_i}\Gamma(1-b_{ji}-\beta_{ji}s_i)\prod_{j=n+1}^{p_i}\Gamma(a_{ji},\alpha_{ji}s_i)\right\}} \tag{6.1.4}$$

For the detailed study of I-function, we refer the original paper of Saxena (1982).

The general class of polynomials defined by Srivastava [(1973), p.1, (1)] are represented in the following manner:

$$S_{\mu}^{\nu}[x]=\sum_{k=0}^{[\mu/\nu]}\frac{(-\mu)_{\nu t}}{t!}A_{\mu,t}x^{t}, \quad \mu=0,1,2... \tag{6.1.5}$$

Where ν an arbitrary positive integer and the coefficients $A_{\mu,\nu}(\mu,\nu\geq 0)$ are arbitrary constant, real or complex.

Let $f(x)\in\phi$; *where* ϕ denotes the class of functions for which

$$f(x)=\begin{cases} 0(x^{\eta}) & for\, small\, values\, of\, x \\ 0\left(x^{-\epsilon}e^{-\xi x}\right) & for\, large\, values\, of\, x \end{cases} \tag{6.1.6}$$

On considering the asymptotic expansion of the I-function [Saxena (1982)], it is found that the operators defined by (1.1) and (1.2) exist under the following sets of conditions:

$$\begin{cases} \sum_{j=1}^{n}\alpha_j-\sum_{j=n+1}^{p_i}\alpha_{ji}+\sum_{j=1}^{m}\beta_j-\sum_{j=m+1}^{q_i}\beta_{ji}\equiv\delta>0; \\ |\arg z|<\frac{1}{2}\pi\delta; \quad \sum_{j=1}^{q_i}\beta_{ji}-\sum_{j=1}^{p_i}\alpha_{ji}\geq 0 \end{cases} \tag{6.1.7}$$

$$\begin{cases} \sum_{j=1}^{n'}\alpha'_j-\sum_{j=n'+1}^{p'_i}\alpha'_{ji}+\sum_{j=1}^{m'}\beta'_j-\sum_{j=m'+1}^{q'_i}\beta'_{ji}\equiv\delta'>0; \\ |\arg z'|<\frac{1}{2}\pi\delta'; \quad \sum_{j=1}^{q'_i}\beta'_{ji}-\sum_{j=1}^{p'_i}\alpha'_{ji}\geq 0 \end{cases} \tag{6.1.8}$$

and y, y' are suitably bounded complex variables.

The following results are needed in the sequel [Gradsteyn,Ryzhik (1980), p.286, (3.197,3)], [Erdelyi et.al(1953),p.64,(23)], [Erdelyi et.al(1953),p.201,(8)] and [Erdelyi et.al(1953), p.62,(15)].

$$\int_0^1 x^{\lambda-1}(1-x)^{\mu-1}(1-\beta x)^{-\nu}\,dx = B(\lambda,\mu)\,{}_2F_1(\nu,\lambda;\lambda+\mu;\beta) \tag{6.1.9}$$

Provided that $\text{Re}(\lambda)>0,\ \text{Re}(\mu)>0, |\beta|<1$.

$${}_2F_1(a,b;c;z) = (1-z)^{c-a-b}\,{}_2F_1(c-a,c-b;c;z) \tag{6.1.10}$$

Provided that $|z|<1$.

$$\int_y^\infty x^{-\lambda}(x+\alpha)^{\nu}(x-y)^{\mu-1}dx = y^{\mu+\nu-\lambda}B(\lambda,\lambda-\mu-\nu)\left(1-\frac{\alpha}{y}\right)^{\mu+\nu}{}_2F_1\left(\lambda,\mu;\lambda-\mu;-\frac{\alpha}{y}\right) \tag{6.1.11}$$

Provided that $0<\text{Re}(\mu)<\text{Re}(\lambda-\mu), \left|\frac{\alpha}{y}\right|<1$.

$${}_2F_1(a,b;c;-z) = \frac{1}{2\pi\omega}\int_{c-\omega\infty}^{c+\omega\infty}\frac{\Gamma(a+s)\Gamma(b+s)}{\Gamma(c+s)}\Gamma(-s)\,z^s\,ds \tag{6.1.12}$$

Provided that $|\arg(-z)|<\pi$ *and* $\omega=\sqrt{-1}$.

We shall use the following notations:

$$A^* = (a_j,\alpha_j)_{1,n},(a_{ji},\alpha_{ji})_{n+1,p_i}; B^* = (b_j,\beta_j)_{1,m},(b_{ji},\beta_{ji})_{m+1,q_i}$$
$$C^* = (a'_j,\alpha'_j)_{1,n'},(a'_{ji},\alpha'_{ji})_{n'+1,p'_i}; D^* = (b'_j,\beta'_j)_{1,m'},(b'_{ji},\beta'_{ji})_{m'+1,q'_i}$$

2. Compositions of the operators defined by (6.1.1) and (6.1.2):

Theorem 1:

If $f(x)\in\phi; \text{Re}(\lambda+\eta)>-1, \text{Re}(h-\lambda)>0, \text{Re}\left(\beta+\sigma t+c\frac{b_j}{\beta_j}\right)>0;$

$\text{Re}\left(\beta+\sigma t'+c\frac{b_j}{\beta_j}\right)>0; j=1,\dots,m; t,t'=0,1,2,\dots,\frac{b}{a}$ then under the conditions (6.1.7), the following result holds

$$Y_{\alpha}^{h} Y_{\beta}^{\lambda}\{f(x)\} = \frac{x^{-\alpha-\beta-\lambda}}{\Gamma(\alpha)\Gamma\beta)} \sum_{t,t'}^{[b/a]} \frac{(-b)_{at}(-b)_{at'}}{t!t'!} A_{b,t} A_{b,t'} y^{t+t'} x^{-\sigma(t+t')}$$

$$\int_0^x u^{\lambda}(x-u)^{\alpha+\beta+\sigma(t+t')-1} I^{0,2:m,n\;;m,n+1\;\;1,0}_{2,1:p_i,q_i:r;p_i+1,q_i+1:r;0,1} \left[\begin{matrix} z\left(1-u/x\right)^c \\ z\left(1-u/x\right) \\ \left(\frac{u}{x}-1\right) \end{matrix} \left| \begin{matrix} (1-\lambda-\beta-\sigma t'+h;0,c,1),(1-\alpha-\sigma t;c,0,1);A^*,(1-\beta-\sigma t;c);A^* \\ (1-\alpha-\beta-\sigma(t+t');c,c,1),B^*;B^*,(1-\lambda-\beta-\sigma t'+\lambda,c);(0,1) \end{matrix} \right. \right] f(u)\,du$$

(6.1.13)

Where $I[\mathrm{x,y,z}]$ appearing on the R.H.S. of (6.1.13) is the I-function

of three variables defined and represented as follows:

$$I\begin{bmatrix} x \\ y \\ z \end{bmatrix} = I^{0,n;\; m_1,n_1\; :m_2,n_2\; :m_3,n_3}_{p_i,q_i:r;p_{i1},q_{i1},:r;p_{i2},q_{i2},:r;p_{i3},q_{i3}:r}$$

$$\left[\begin{matrix} z_1 \\ z_2 \\ z_3 \end{matrix} \left| \begin{matrix} (a_j;A_J',A_J'',A_J''')_{1,p_i};(\tau_j',C_j')_{1,m_1},(\tau_{ji1}',C_{ji1}')_{m_1+1,p_{i1}};(\tau_j'',C_j'')_{1,n_2},(\tau_{ji2}'',C_{ji2}'')_{n_2+1,p_{i2}};(\tau_j''',C_j''')_{1,n_3},(\tau_{ji3}''',C_{ji3}''')_{n+1,p_{i3}} \\ (b_j;B_J',B_J'',B_J''')_{1,q_i};(d_j',D_j')_{1,m_1},(d_{ji1}',D_{ji1}')_{m_1+1,q_{i1}};(d_j'',D_j'')_{1,m_2},(d_{ji2}'',D_{ji2}'')_{m_2+1,q_{i2}};(d_j''',D_j''')_{1,m_3},(d_{ji3}''',D_{ji3}''')_{m_3+1,q_{i3}} \end{matrix} \right. \right]$$

(6.1.14)

$$\phi(s_1,s_2,s_3) = \frac{\prod_{j=1}^{n}\Gamma\left(1-a_j+\sum_{j=1}^{3}A_j^k s_k\right)}{\prod_{j=n+1}^{p_i}\Gamma\left(a_{ji}-\sum_{j=1}^{3}A_{ji}^k s_k\right)\prod_{j=1}^{q_i}\Gamma\left(1-b_{ji}+\sum_{j=1}^{3}B_{ji}^k s_k\right)} \qquad (6.1.15)$$

$$\theta_k(s_k) = \frac{\prod_{j=1}^{m_k}\Gamma\left(d_j^k-D_j^k s_k\right)\prod_{j=1}^{n_k}\Gamma\left(1-\tau_j^k+C_j^k s_k\right)}{\sum_{i=1}^{r}\left\{\prod_{j=m_{ik}+1}^{q_{ik}}\Gamma\left(1-d_{ji}^k+D_{ji}^k s_k\right)\prod_{j=n_k+1}^{p_{ik}}\Gamma\left(\tau_{ji}^k-C_{ji}^k s_k\right)\right\}} \qquad (6.1.16)$$

$\forall\; k \in \{1,2,3\}$

The triple Mellin-Bernes contour integral representing the I-function of three variables.

(6.1.14) converges absolutely under the following conditions:

$$\Lambda_k = \sum_{j=1}^{p_i}A_{ji}^k + \sum_{j=1}^{p_{ik}}C_{ji}^k - \sum_{j=1}^{q_i}B_j^k - \sum_{j=1}^{q_{ik}}D_{ji}^k \le 0; \quad \forall\; k \in \{1,2,3\} \qquad (6.1.17)$$

$$\Omega_k = -\sum_{j=n+1}^{p_i} A_j^k + \sum_{j=1}^{n_k} C_j^k - \sum_{j=n_k+1}^{p_{ik}} C_{ji}^k - \sum_{j=1}^{q_i} B_j^k + \sum_{j=1}^{m_k} D_j^k - \sum_{j=m_k+1}^{q_{ik}} D_{ji}^k > 0; \forall \; k \in \{1,2,3\}$$

(6.1.18)

And $|\arg(z_k)| < \frac{1}{2}\pi\Omega_k \;; \forall \; k \in \{1,2,3\}$

Proof: To prove (6.1.13), we employ the definition (6.1.1), and then it follows that

$$Y_\alpha^h Y_\beta^\lambda \{f(x)\} = \frac{x^{-h-\alpha}}{\Gamma(\alpha)} \int_0^x (x-s)^{\alpha-1} I_{p_i,q_i:r}^{m,n}\left[z\left(1-\frac{s}{x}\right)^c\right] S_b^a\left[y\left(1-\frac{s}{x}\right)^\sigma\right] s^{-\lambda-\beta+h}$$
$$\left\{\frac{1}{\Gamma(s)}\int_0^s (s-u)^{\beta-1} I_{p_i,q_i:r}^{m,n}\left[z\left(1-\frac{u}{s}\right)^c\right] S_b^a\left[y\left(1-\frac{u}{s}\right)^\sigma\right] u^\lambda f(u)\,du\right\} ds \qquad (6.1.19)$$

If we interchange the order of integration, which is justified under the conditions stated with the theorem, we obtain

$$Y_\alpha^h Y_\beta^\lambda \{f(x)\} = \frac{x^{-h-\alpha}}{\Gamma(\alpha)\Gamma(\beta)} \int_0^x \Omega\, u^\lambda f(u)\,du \qquad (6.1.20)$$

Where

$$\Omega = \int_u^x (x-s)^{\alpha-1} s^{h-\lambda-\beta} (s-u)^{\beta-1} I_{p_i,q_i:r}^{m,n}\left[z\left(1-\frac{s}{x}\right)^c\right] S_b^a\left[y\left(1-\frac{s}{x}\right)^\sigma\right]$$
$$I_{p_i,q_i:r}^{m,n}\left[z\left(1-\frac{u}{s}\right)^c\right] S_b^a\left[y\left(1-\frac{u}{s}\right)^\sigma\right] \qquad (6.1.21)$$

If we write the series representation (6.1.5) for $S_\mu^\nu[z]$ and Mellin-Barnes integrals [Srivastava,Gupta and Goyal (1982), p.10, (2.1.1)] for the I-function appearing in (6.1.21) and interchange the order of integration, then the s-integral can be evaluated by setting $\omega = \frac{x-s}{x-u}$ and using the result (6.1.9).

Applying the result (6.1.12) for ${}_2F_1(.)$ occurring in the resulting expression and interpreting the result thus obtained with the help of (6.1.14), we find that

$$\Omega = \sum_{t,t'=0}^{[b/a]} \frac{(-b)_{at}(-b)_{at'}}{t!\,t'!} A_{b,t} A_{b,t'} y^{t+t'} x^{-\sigma(t+t')+h-\lambda-\beta} (x-u)^{\alpha+\beta+\sigma(t+t')-1}$$

$$I^{0,2:m,n\ :m,n+1\ :1,0}_{2,1:p_i,q_i:r:p_i+1,q_i+1:r:0,1}\left[\begin{matrix} z\left(1-u/x\right)^c \\ z\left(1-u/x\right)^c \\ \left(u/x-1\right) \end{matrix}\left|\begin{matrix}(1-\lambda-\beta-\sigma t'+h;0,c,1),(1-\alpha-\sigma t;c,0,1);A^*;- \\ (1-\alpha-\beta-\sigma(t+t');c,c,1),-,B^*,B^*,(1-\lambda-\beta-\sigma t'+h,c);(0,1)\end{matrix}\right.\right]$$

If we substitute the value of Ω in (6.1.20) then we arrive at the result (6.1.13).

The following theorem 2 can be proved in a similar manner by employing the definition

(6.1.2) and the substitution $\omega = \dfrac{(u-s)}{(u-x)}$

Theorem 2:

If $f(x) \in \phi, \mathrm{Re}(\lambda - h) > 0, \mathrm{Re}(\xi + \lambda) > 0, \mathrm{Re}\left[\alpha + \rho t + d\, b'_j / \beta'_j\right] > 0,$

$\mathrm{Re}\left[\beta + \rho t' + d\, b'_j / \beta'_j\right] > 0; \forall\, j = 1,\ldots,m'; t,t' = 0,1,\ldots,\left[f/e\right]$, *then under the conditions* (1.8) *the following result holds:*

$$V_\alpha^h V_\beta^\lambda \{f(x)\} = \frac{x^h}{\Gamma(\alpha)\Gamma(\beta)} \sum_{t,t'=0}^{[f/e]} \frac{(-f)_{et}(-f)_{et'}}{t!\,t'!} A_{f,t} A_{f,t'} (y')^{t+t'} \int_x^\infty u^{-\alpha-\beta-h-\rho(t+t')}$$

$$(u-x)^{\alpha+\beta+\rho(t+t')-1} I^{0,2:m',n'\ ;m',n'\ ;1,0}_{2,1:p'_i+1,q'_i+1:r;p'_i,q'_i:r;0,1}\left[\begin{matrix} z'\left(1-\frac{x}{u}\right)^d \\ z'\left(1-\frac{u}{x}\right)^d \\ \left(\frac{u}{x}-1\right) \end{matrix}\left|\begin{matrix}(1-h-\alpha-\rho t+\lambda;d,0,1),(1-\beta-\rho t';0,d,1),(1-\alpha-\rho t,d);C^*,C^* \\ (1-\alpha-\beta-\rho(t+t');d,d,1);-\ ;-\ ;D^*;D^*\end{matrix}\right.\right] f(u)\,du \tag{6.1.22}$$

3. Composition of a mixed type:

Theorem3:

$$If\ f(x)\in\phi;\min\{\mathrm{Re}(h+\xi+1),\mathrm{Re}(h+\eta+1)\}>0;\mathrm{Re}(\lambda+\xi)>0;\mathrm{Re}\left(\beta+\rho t'+d\frac{b_j'}{\beta_j'}\right)>0;$$

$$\mathrm{Re}\left(\alpha+\sigma t+c\frac{b_k}{\beta_k}\right)>0;\ j=1,\ldots,m';k=1,\ldots,m;t=0,1,\ldots,\left[\frac{b}{a}\right];t'=0,1,\ldots,\left[\frac{f}{e}\right];$$

Then under the conditions (6.1.7) and (6.1.8), the following result holds:

$$Y_\alpha^h V_\beta^\lambda\{f(x)\}=V_\beta^\lambda Y_\alpha^h\{f(x)\}=\frac{x^{-1-h}}{\Gamma(\alpha)\Gamma(\beta)}\Gamma(1+\lambda+\beta)\sum_{t=0}^{[b/a]}\sum_{t'=0}^{[f/e]}\frac{(-b)_{at}(-f)_{et'}}{t!t'!}A_{b,t}A_{f,t'}y^t(y')^{t'}$$

$$\int_0^x u^h\left(1-\frac{u}{x}\right)^{\alpha+\beta+\sigma t-\rho t'-1} I^{0,2:m,n\ ;m',n'\ ;1,0}_{2,2:p_i,q_i:r;p_i',q_i':r;0,1}\left[\begin{matrix} z\left(1-\frac{u}{x}\right)^c \\ z'\left(1-\frac{u}{x}\right)^d \\ \left(-\frac{u}{x}\right)\end{matrix}\left|\begin{matrix}(1-\lambda-\beta-h-\alpha-\sigma t-\rho t';c,d,1),(1-\beta-\rho t';0,d,1);A^*;C^*\\(1-\lambda-\beta-h-\alpha-\sigma t-\rho t';c,d,0),(-\lambda-\beta-h-\rho t';0,d,1);B^*;D^*\end{matrix}\right.\right]f(u)\,du$$

$$+\frac{x^\lambda}{\Gamma(\alpha)\Gamma(\beta)}\Gamma(1+\lambda+h)\sum_{t=0}^{[b/a]}\sum_{t'=0}^{[f/e]}\frac{(-b)_{at}(-f)_{et'}}{t!t'!}A_{b,t}A_{f,t'}y^t(y')^{t'}$$

$$\int_x^\infty u^{-\lambda-1}\left(1-\frac{x}{u}\right)^{\alpha+\beta+\sigma t-\rho t'-1} I^{0,2:m,n\ ;m',n'}_{2,2:p_i,q_i:r;p_i',q_i':r}\left[\begin{matrix} z\left(1-\frac{x}{u}\right)^c \\ z'\left(1-\frac{x}{u}\right)^d \\ \left(-\frac{x}{u}\right)\end{matrix}\left|\begin{matrix}(1-\lambda-\beta-h-\alpha-\sigma t-\rho t';c,d,1),(1-\alpha-\sigma t;c,0,1);A^*;C^*\\(1-\lambda-\beta-h-\alpha-\sigma t-\rho t';c,d,0),(-\lambda-\alpha-h-\sigma t;c,0,1);B^*;D^*\end{matrix}\right.\right]f(u)\,du$$

(6.1.23)

Proof: From (6.1.1) and (6.1.2), it follows that

$$Y_\alpha^h V_\beta^\lambda\{f(x)\}=\frac{x^{-h-\alpha}}{\Gamma(\alpha)\Gamma(\beta)}\int_0^x(x-s)^{\alpha-1}I^{m,n}_{p_i,q_i:r}\left[z\left(1-\frac{s}{x}\right)^c\right]S_b^a\left[y\left(1-\frac{s}{x}\right)^\sigma\right]$$

$$\left\{s^{h+\alpha}\int_x^\infty(u-s)^{\beta-1}u^{-\lambda-\beta}I^{m',n'}_{p_i',q_i':r}\left[z'\left(1-\frac{s}{x}\right)^c\right]S_f^e\left[y'\left(1-\frac{s}{x}\right)^\sigma\right]f(u)\,du\right\}ds$$

(6.1.24)

On interchanging the order of integration, which is justified in view of the conditions with the theorem, and using Fubini's theorem, it is seen that

$$Y_{\alpha}^{h}\, V_{\beta}^{\lambda}\{f(x)\} = \int_{0}^{\infty} \psi(x,u)\, f(u)\, du \qquad (6.1.25)$$

Where

$$\psi(x,u) = \frac{x^{-h-\alpha} u^{-\lambda-\beta}}{\Gamma(\alpha)\Gamma(\beta)} \int_{0}^{\min(x,u)} (x-s)^{\alpha-1}(u-s)^{\beta-1} s^{h+\lambda} I_{p_i,q_i:r}^{m,n}\left[z\left(1-\frac{s}{x}\right)^{c}\right] S_{b}^{a}\left[y\left(1-\frac{s}{x}\right)^{\sigma}\right]$$
$$I_{p_i',q_i':r'}^{m',n'}\left[z'\left(1-\frac{s}{x}\right)^{d}\right] S_{f}^{e}\left[y'\left(1-\frac{s}{x}\right)^{\rho}\right] ds$$

(6.1.26)

The integral (6.1.25) can be evaluated under the following two cases:

(i) $x > u$ and (ii) $x < u$

The other composition of a mixed type can be obtained similarly, and we finally obtain

$$V_{\beta}^{\lambda}\, Y_{\alpha}^{h}\{f(x)\} = \frac{x^{\lambda}}{\Gamma(\alpha)\Gamma(\beta)}\left[\int_{0}^{x} u^{h}\, \Omega' f(u)\, du + \int_{x}^{\infty} u^{h} \Lambda' f(u)\, du \right] \qquad (6.1.27)$$

where

$$\Omega' = \int_{x}^{\infty} (s-x)^{\beta-1} s^{-\lambda-\beta-h-\alpha} (s-u)^{\alpha-1} I_{p_i,q_i:r}^{m,n}\left[z\left(1-\frac{u}{s}\right)^{c}\right] S_{b}^{a}\left[y\left(1-\frac{u}{s}\right)^{\sigma}\right]$$
$$I_{p_i',q_i':r'}^{m',n'}\left[z'\left(1-\frac{x}{s}\right)^{d}\right] S_{f}^{e}\left[y'\left(1-\frac{u}{s}\right)^{\rho}\right] ds \qquad (6.1.28)$$

and

$$\Lambda' = \int_{u}^{\infty} (s-x)^{\beta-1} s^{-\lambda-\beta-h-\alpha} (s-u)^{\alpha-1} I_{p_i,q_i:r}^{m,n}\left[z\left(1-\frac{u}{s}\right)^{c}\right] S_{b}^{a}\left[y\left(1-\frac{u}{s}\right)^{\sigma}\right]$$
$$I_{p_i',q_i':r'}^{m',n'}\left[z'\left(1-\frac{x}{s}\right)^{d}\right] S_{f}^{e}\left[y'\left(1-\frac{x}{s}\right)^{\rho}\right] ds \qquad (6.1.29)$$

To determine the value of Ω' *and* Λ', we express $S_{\mu}^{\nu}[z]$ in terms of its equivalent series (6.1.5) and the I-function as Mellin-Bernes integrals [Srivastava, Gupta and Goyal (1982),p.10, (2.1.1)] and interchange the order of integration, which is justified under the conditions stated with the theorem.

If we evaluate the s -integral with the help of the result (6.1.11) and apply (6.1.12) and interpret the results thus obtained with the help of (6.1.14), then the values of Ω' and Λ' respectively are obtained as follows

$$\Omega' = x^{-1-\lambda-h}\Gamma(1+\lambda+h)\sum_{t=0}^{[b/a]}\sum_{t'=0}^{[f/e]}\frac{(-b)_{at}(f)_{et'}}{t!\,t'!}A_{b,t}\,A_{f,t'}\,y^{t}(y)^{t'}\,I^{0,2:m,n\;;m',n'\;;1,0}_{2,2:p_i,q_i:r;p'_i,q'_i:r;0,1}\left[\begin{matrix} z\left(1-\frac{u}{x}\right)^{c} \\ z'\left(1-\frac{u}{x}\right)^{d} \\ \left(\frac{u}{x}\right)\end{matrix}\middle|\begin{matrix}(1-\lambda-\beta-h-\alpha-\sigma t-\rho t';c,d,1),(1-\alpha-\sigma t;o,d,1);A^{*};C^{*} \\ (1-\lambda-\beta-h-\alpha-\sigma t-\rho t';c,d,0),(-\lambda-\alpha-h-\sigma t;0,d,1);B^{*};D^{*};(0,1)\end{matrix}\right] \tag{6.1.30}$$

and

$$\Lambda' = \Gamma(1+\lambda+h)\sum_{t=0}^{[b/a]}\sum_{t'=0}^{[f/e]}\frac{(-b)_{at}(f)_{et'}}{t!\,t'!}A_{b,t}\,A_{f,t'}\,y^{t}(y)^{t'}\,u^{-\lambda-h-1}\left(1-\frac{u}{x}\right)^{\beta+\rho t'+\alpha+\sigma t-1}$$

$$I^{0,2:m,n\;;m',n'\;;1,0}_{2,2:p_i,q_i:r;p'_i,q'_i:r;0,1}\left[\begin{matrix} z\left(1-\frac{u}{x}\right)^{c} \\ z'\left(1-\frac{u}{x}\right)^{d} \\ \left(\frac{u}{x}\right)\end{matrix}\middle|\begin{matrix}(1-\lambda-\beta-h-\alpha-\sigma t-\rho t';c,d,1),(1-\alpha-\sigma t;c,0,1);A^{*};C^{*} \\ (1-\lambda-\beta-h-\alpha-\sigma t-\rho t';c,d,0),(-\lambda-\alpha-h-\sigma t;c,0,1);B^{*};D^{*};(0,1)\end{matrix}\right] \tag{6.1.31}$$

The values of Ω' and Λ' when substituted in (6.1.27) give rise to the result (6.1.23).

4. Varification of commutativity of Y_{α}^{h} and V_{β}^{λ} :

With some obvious parametric interchange (6.1.13) can be written as

$$Y_{\beta}^{\lambda}\,Y_{\alpha}^{h}\left\{F(x)\right\}=\frac{x^{-\alpha-\beta}}{\Gamma(\alpha)\Gamma(\beta)}\sum_{t,t'=0}^{\left[b/a\right]}\frac{(-b)_{at}(-b)_{at'}}{t!\,t'!}A_{b,t}\,A_{b,t'}\,y^{t+t'}\,x^{-\sigma(t+t')}$$

$$\int_{0}^{x}(x-u)^{\alpha+\beta=\sigma(t+t')-1}\left(\frac{u}{x}\right)^{h} I_{2,1:p_i,q_i:r;p_i+1,q_i+1:r:0,1}^{0,2;m,n;\quad m,n+1;\quad 1,0}\left[\begin{matrix} z\left(1-\frac{u}{x}\right)^{c} \\ z\left(1-\frac{u}{x}\right)^{c} \\ \left(\frac{u}{x}-1\right)\end{matrix}\left|\begin{matrix}(1-h-\alpha-\sigma t+\lambda;0,c,1),(1-\beta-\sigma t';c,0,1),A^{*}(1-\alpha-\sigma t;c),A^{*} \\ (1-\alpha-\beta-\sigma(t+t');c,c,1);B^{*};B^{*};(0,1)\end{matrix}\right.\right]$$

(6.1.32)

If we express the I-function of three variables in terms of Mellin-Bernes integral (6.1.13) and apply the result (6.1.12), we then obtain

$$Y_{\beta}^{\lambda}\,Y_{\alpha}^{h}\left\{F(x)\right\}=\frac{x^{-\alpha-\beta}}{\Gamma(\alpha)\Gamma(\beta)}\sum_{t,t'=0}^{\left[b/a\right]}\frac{(-b)_{at}(-b)_{at'}}{t!\,t'!}A_{b,t}\,A_{b,t'}\,y^{t+t'}\,x^{-\sigma(t+t')} \tag{6.1.33}$$

By the application of Euler's transformation of the hypergeometric function (6.1.10), the result (6.1.33) transforms into

$$\frac{x^{-\alpha-\beta}}{\Gamma(\alpha)\Gamma(\beta)}\sum_{t,t'=0}^{\left[b/a\right]}\frac{(-b)_{at}(-b)_{at'}}{t!\,t'!}A_{b,t}\,A_{b,t'}\,y^{t+t'}\,x^{-\sigma(t+t')}\int_{0}^{x}(x-u)^{\alpha+\beta+\sigma(t+t')-1}u^{\lambda}\,I_{2,1:p_i,q_i:r;p_i+1,q_i+1:r:0,1}^{0,2;m,n;\quad m,n+1;\quad 1,0}$$

$$\left[\begin{matrix} z\left(1-\frac{u}{x}\right)^{c} \\ z\left(1-\frac{u}{x}\right)^{c} \\ \left(\frac{u}{x}-1\right)\end{matrix}\left|\begin{matrix}(1-\lambda-\beta-\sigma t'+h;0,c,1),(1-\alpha-\sigma t;c,0,1),A^{*},(1-\beta-\sigma t';c),A^{*} \\ (1-\alpha-\beta-\sigma(t+t');c,c,1);B^{*};B^{*},(1-\lambda-\beta-\sigma t'+h,c);(0,1)\end{matrix}\right.\right]f(u)\,du=Y_{\alpha}^{h}\,Y_{\beta}^{\lambda}\left\{f(x)\right\} \tag{6.1.34}$$

The result (6.1.34) shows that the operator Y_{α}^{h} and Y_{β}^{λ} commute.

Similarly, it is easy to verify the commutativity of the fractional integral operators involved in (6.1.2).

5. Special Cases:

For $\mathrm{r}=1$, the I-function appearing in (6.1.1) and (6.1.2) reduce to the corresponding Fox's H-function and consequently we obtain the following operators associated with H-function

$$R_{\alpha}^{h}\{f(x)\}=x^{-h-\alpha}A_{\alpha}\{x^{h}f(x)\}=\frac{x^{-h-\alpha}}{\Gamma(\alpha)}\int_{0}^{x}(x-s)^{\alpha-1}H_{p_1,q_1:r}^{m,n}\left[z\left(1-\frac{s}{x}\right)^{c}\right]$$

$$S_{b}^{a}\left[y\left(1-\frac{s}{x}\right)^{\sigma}\right]s^{h}f(s)\,ds \qquad (6.1.35)$$

and

$$K_{\beta}^{\lambda}\{f(x)\}=x^{\lambda}B_{\beta}\{x^{-\lambda-\beta}f(x)\}=\frac{x^{\lambda}}{\Gamma(\beta)}\int_{x}^{\infty}(s-x)^{\beta-1}H_{p_1^{'},q_1^{'}:r}^{m^{'},n^{'}}\left[z^{'}\left(1-\frac{x}{s}\right)^{d}\right]$$

$$S_{b}^{a}\left[y\left(1-\frac{s}{x}\right)^{\sigma}\right]s^{h}f(s)\,ds \qquad (6.1.36)$$

The operators (6.1.35) and (6.1.36) are valid under the following sets of conditions:

$(i)\, f(x)\in\phi$

$(ii)\,\mathrm{Re}(h+\eta)>-1,\mathrm{Re}(\lambda+\xi)>0;\mathrm{Re}\left(\alpha+\sigma t+c\frac{b_j}{\beta_j}\right)>0;\mathrm{Re}\left(\beta+\rho t^{'}+d\frac{b_k^{'}}{\beta_k^{'}}\right)>0;$

$j=1,\ldots,m;k=1,\ldots,m^{'},t=0,1,\ldots,\left[b/a\right],t^{'}=0,1,\ldots,\left[f/e\right]$

$$(iii)\sum_{j=1}^{n^{'}}\alpha_{j}^{'}-\sum_{j=n+1}^{p_1^{'}}\alpha_{j}^{'}+\sum_{j=1}^{m^{'}}\beta_{j}^{'}-\sum_{j=m+1}^{q_1^{'}}\beta_{j}^{'}\equiv\delta^{'}>0;$$

$$|\arg z|<\frac{1}{2}\pi\delta^{'};\ \sum_{j=1}^{q_1^{'}}\beta_{j}^{'}-\sum_{j=1}^{p_1^{'}}\alpha_{j}^{'}\geq 0$$

(iv) $$\sum_{j=1}^{n^{'}}\alpha_{j}^{'}-\sum_{j=n+1}^{p_1^{'}}\alpha_{j}^{'}+\sum_{j=1}^{m^{'}}\beta_{j}^{'}-\sum_{j=m+1}^{q_1^{'}}\beta_{j}^{'}\equiv\delta^{'}>0;$$

$$|\arg z|<\frac{1}{2}\pi\delta^{'};\ \sum_{j=1}^{q_1^{'}}\beta_{j}^{'}-\sum_{j=1}^{p_1^{'}}\alpha_{j}^{'}\geq 0$$

and y, y' are suitably bounded complex variables.

Finally it is interesting to observe that the theorem 1, 2 and 3 provide the generalizations of the results given earlier by Saigo (1984), Goyal et. al. (1991), Srivastava and Buschman (1973), Srivastava et. al. (1990), Saxena et. al. (1993) etc..

SECTION 2

INTEGRALS AND FOURIER SERIES INVOLVING GENERALIZED HYPERGEOMETRIC FUNCTIONS

In the present chapter we have evaluated an integral involving an exponential function, Sine function, generalized hypergeometric series and I-function, and we have employed it to evaluate a double integral and establish Fourier series for the product of generalized hypergeometric functions. We have also derived a double Fourier exponential series for the I-function. Our results are unified in nature and act as a key formula from which we can derive many results as their particular cases, but due to lack of space we do not include them here.

1. Introduction :

The I-function occurring in the present chapter is defined and represented as follows [Saxena (1982)] :

$$I[Z] = I^{m,n}_{p_i,q_i;r}[Z] = I^{m,n}_{p_i,q_i;r}\left[z \left| \begin{matrix} (a_j,\alpha_j)_{1,n},(a_{ji},\alpha_{ji})_{n+1,p_i} \\ (b_j,\beta_j)_{1,m},(b_{ji},\beta_{ji})_{m+1,q_i} \end{matrix} \right. \right] = \frac{1}{2\pi\omega}\int_L \chi(\xi) z^{\xi} d\xi \tag{6.2.1}$$

Where $\omega = \sqrt{-1}.$

$$\chi(\xi) = \frac{\prod_{j=1}^{m}\Gamma(b_j-\beta_j\xi)\prod_{j=1}^{n}\Gamma(1-a_j+\alpha_j\xi)}{\sum_{i=1}^{r}\{\prod_{j=m+1}^{q_i}\Gamma(1-b_{ji}+\beta_{ji}\xi)\prod_{j=n+1}^{p_i}\Gamma(a_{ji}-\alpha_{ji}\xi)\}} \tag{6.2.2}$$

$p_i,q_i(i=1,2....,r),m,n$ are integers satisfying $0 \le n \le p_i, 0 \le m \le q_i, r$ is finite, $\alpha_j,\beta_j,\alpha_{ji},\beta_{ji}$ are real and a_j,b_j,a_{ji},b_{ji} are complex numbers such that $\alpha_j(b_h+\nu) \ne \beta_h(\alpha_j-\nu-k)$, for $\nu,k = 0,1,2,....$

We shall use the following notations:

$$A^* = (a_j,\alpha_j)_{1,n},(a_{ji},\alpha_{ji})_{n+1,p_i} \qquad B^* = (b_j,\beta_j)_{1,m},(b_{ji},\beta_{ji})_{m+1,q_i}$$

The following formulae are required in the proofs:

$$\int_0^{\pi}(\sin x)^{\omega-1}e^{imx}\,{}_PF_Q[{}^{\alpha_P}_{\beta_Q};c(\sin x)^{2h}]dx = \frac{\pi e^{im\pi/2}}{2^{\omega-1}}\sum_{r=0}^{\infty}\frac{(\alpha_P)_r c^r\Gamma(\omega+2hr)}{(\beta_Q)_r r! 2^{2hr}\Gamma(\frac{\omega+2hr\pm m+1}{2})} \tag{6.2.3}$$

Where α_P denotes $\alpha_1,...,\alpha_p$; $\Gamma(a \pm b)$ represents $\Gamma(a+b), \Gamma(a-b)$; h is a positive integer; $P < Q(or P = Q+1 and |c| < 1)$; no one of the β_Q is zero, negative integer and $\mathrm{Re}(\omega) > 0$.

$$\int_0^\pi (\sin x)^{\omega-1} e^{imx} {}_P F_Q[{}^{\alpha_P}_{\beta_Q} ; c(\sin x)^{2h}]_U F_V[{}^{\delta_P}_{\gamma_Q} ; d(\sin x)^{2k}] dx =$$

$$\frac{\pi e^{im\pi/2}}{2^{\omega-1}} \sum_{r,t=0}^{\infty} \frac{(\alpha_P)_r c^r (\gamma_U)_t d^t \Gamma(\omega+2hr+2kt)}{(\beta_Q)_r r! (\delta_V)_t t! 2^{2hr+2kt} \Gamma(\frac{\omega+2hr+2kt \pm m+1}{2})} \quad (6.2.4)$$

Where in addition to the condition and notation of (6.2.4), k is a positive integer; $U < V(or U + V + 1 and |d| < 1)$; no one of δ_V is zero or a negative integer.

The following orthogonality properties [Erdelyi et. al (1953)], which can be established easily:

$$\int_0^\pi e^{i(m-n)x} dx = \begin{cases} \pi & m = n \\ \pi & m = n = 0 \\ 0 & m \neq n \end{cases} \quad (6.2.5)$$

$$\int_0^\pi e^{i(m-n)x} \cos nx dx = \begin{cases} \pi/2 & m = n \\ \pi & m = n = 0 \\ 0 & m \neq n \end{cases} \quad (6.2.6)$$

$$\int_0^\pi e^{i(m-n)x} \sin nx dx = \begin{matrix} i\pi/2, m = n \\ 0, m \neq n \end{matrix} \quad (6.2.7)$$

Provided either both m and n are odd or both m and n are even integers.

In what follows for sake of brevity in addition to the notations earlier given in this section , λ and μ are positive numbers and symbol $(a \pm b, \lambda)$ represent $(a+b, \lambda), (a-b, \lambda)$ and

$$I(x) = I_{p_i, q_i; r}^{U,V} \left[z(\sin x)^{-2\lambda} \middle| {}^{A^*}_{B^*} \right] \phi(r) = \frac{(\alpha_P)_r c^r}{(\beta_Q)_r r!} ; \varphi(t) = \frac{(\gamma_U)_r d^r}{(\delta_V)_t t!}$$

$$I[x,y] = I^{U,V}_{p_i,q_i;r}\left[z(\sin x)^{-2\lambda}(\sin y)^{-2\mu}\left|\begin{matrix}A^*\\B^*\end{matrix}\right.\right]$$

$$F_1(x) = {}_PF_Q\left[\begin{matrix}(\alpha_P)\\(\beta_Q)\end{matrix};c(\sin x)^{2h}\right], F_2(x) = {}_UF_V\left[\begin{matrix}(\gamma_U)\\(\delta_V)\end{matrix};d(\sin x)^{2k}\right]$$

2.Integral:

The integral to be evaluated is

$$\int_0^{\pi}(\sin x)^{\omega-1}e^{imx}F_1(x)F_2(x)I(x)dx$$

$$= \sqrt{(\pi)}e^{im\pi/2}\sum\nolimits_{r,t=0}^{\infty}\phi(r)\psi(t)I^{u+2,v}_{p_i+2,q_i+2;r}\left[z\left|\begin{matrix}A^*,(\frac{\omega+2hr+2kt\pm m+1}{2},\lambda)\\(\frac{\omega+2hr+2kt}{2},\lambda),(\frac{\omega+2hr+2kt+1}{2},\lambda),B^*\end{matrix}\right.\right]$$

(6.2.8)

Provided that

$$\sum\nolimits_{j=1}^{p_i}a_{ji}-\sum\nolimits_{j=1}^{q_i}b_{ji}\le 0, \mathrm{Re}[\omega+2\lambda\frac{(1-a_j)}{b_j}]>0,(j=1,...,u),$$

$$|\arg z|<\frac{1}{2}\pi\Omega, \Omega>0.$$

where

$$\Omega=\sum\nolimits_{j=1}^{m}\beta_j+\sum\nolimits_{j=1}^{n}\alpha_j-\sum\nolimits_{j=m+1}^{q_i}\beta_{ji}-\sum\nolimits_{j=n+1}^{p_i}\alpha_{ji}>0$$

Together with the other conditions given in (6.2.3) and (6.2.4) are satisfied.

Proof: to establish (6.2.8), expressing the I-function in the integrand as the Mellin-Barnes type integral (6.2.1) and changing the order of integrations (which is justified due to the absolute convergence of the integrals in the process), we have

$$\frac{1}{2\pi\omega}\int_L \chi(\xi)z^{\xi}\int_0^{\pi}(\sin x)^{\omega-2\lambda\beta-1}e^{imx}F_1(x)F_2(x)dxd\xi$$

Evaluating the inner integral with the help of (6.2.5) and using multiplication formula for the Gamma-function[Saxena (1982), p.4(11)],we get

$$\sqrt{(\pi)}e^{im\pi/2}\sum_{r,t=0}^{\infty}\phi(t)\psi(t)\frac{1}{2\pi\omega}\int_L \frac{\chi(\xi)\Gamma(\frac{\omega+2hr+2kt}{2}-\lambda\xi)\Gamma(\frac{\omega+2hr+2kt+1}{2}-\lambda\xi)}{\Gamma(\frac{\omega+2hr+2kt\pm m+1}{2}-\lambda\xi)}z^{\xi}d\xi$$

Now applying (6.2.1), the value of the integral (6.2.8) is obtained.

3. Double integral:

The double integral to be evaluated is

$$\int_0^{\pi}\int_0^{\pi}(\sin x)^{\omega_1-1}(\sin y)^{\omega_2-1}e^{i(m_1x+m_2y)}F_1(x)F_2(x)F_1(y)F_2(y)I(xy)dxdy =$$

$$\pi e^{i(m_1+m_2)\pi/2}\sum_{r_1,t_1=0}^{\infty}\sum_{r_2,t_2=0}^{\infty}\phi(r_1)\phi(r_2)\psi(t_1)\psi(t_2).$$

$$I_{p_i+4,q_i+4r}^{u+4,v}\left[z\left|\begin{matrix}A^*,(\frac{\omega_1+2hr_1+2kt_1\pm m_1+1}{2},\lambda),(\frac{\omega_2+2hr_2+2kt_2\pm m_2+1}{2},\mu)\\(\frac{\omega_1+2hr_1+2kt_1}{2},\lambda),(\frac{\omega_1+2hr_1+2kt_1+1}{2},\lambda),(\frac{\omega_2+2hr_2+2kt_2}{2},\mu)(\frac{\omega_2+2hr_2+2kt_2+1}{2},\mu),B^*\end{matrix}\right.\right]$$

(6.2.9)

Where in addition to the conditions stated in (6.2.8).

$$\mathrm{Re}[\omega_1+2\lambda(1-a_j)/\alpha_j]>0,\mathrm{Re}[\omega_2+2\mu(1-a_j)/\alpha_j]>0,(j=1,...,\mu).$$

Proof: To establish (6.2.9), evaluating the x-integral of (6.2.9) with the help of (6.2.8) interchange the order of integration and summation, we get

$$\sqrt{\pi}e^{im_xx/2}\sum_{r_1,r_2=0}^{\infty}\phi(r_1)\phi(r_2)\int_0^{\pi}(\sin y)^{\omega_2-1}e^{im_2y}F_1(y)F_2(y)$$

$$I^{u+2,v}_{p_i+2,q_i+2:r}\left[z(\sin y)^{-2\mu} \left| \begin{matrix} A^*, (\frac{\omega_1+2hr_1+2kt_1\pm m+1}{2},\lambda) \\ (\frac{\omega_1+2hr_1+2kt_1}{2},\lambda),(\frac{\omega_1+2hr_1+2kt_1+1}{2},\lambda),B^* \end{matrix} \right. \right] dy$$

Now applying (6.2.8) to evaluate the y-integral, the value of (6.2.9) is obtained.

4. Fourier Series:

The Fourier series to be established are:

$$(\sin x)^{\omega-1}F_1(x)F_2(x)I(x)=\frac{1}{\sqrt{\pi}}\sum_{n=-\infty}^{\infty}\sum_{t=0}^{\infty}\phi(r)\psi(t)e^{in(\frac{\pi}{2}-x)} \tag{6.2.10}$$

$$(\sin x)^{\omega-1}F_1(x)F_2(x)I(x)=$$

$$\frac{1}{\sqrt{\pi}}\sum_{r,t=0}^{\infty}\phi(r)\psi(t)I^{u+1,v}_{p_i+1,q_i+1:r}\left[z \left| \begin{matrix} A^*, (\frac{\omega+2hr+2kt+1}{2},\lambda) \\ (\frac{\omega+2hr+2kt}{2},\lambda),B^* \end{matrix} \right. \right]+$$

$$\frac{2}{\sqrt{\pi}}\sum_{n=1}^{\infty}\sum_{r,t=0}^{\infty}\phi(r)\psi(t)e^{in\pi/2}I_2(x)\cos nx \tag{6.2.11}$$

$$(\sin x)^{\omega-1}F_1(x)F_2(x)I(x)=\frac{1}{\sqrt{i\pi}}\sum_{n=1}^{\infty}\sum_{r,t=0}^{\infty}\phi(r)\psi(t)e^{in\pi/2}I(x)\sin nx \tag{6.2.12}$$

Where $n's$ are either even or odd in addition to the condition of validity followed by (6.2.8) and

$$I_2(x)=I^{u+2,v}_{p_i+2,q_i+2;2}\left[z \left| \begin{matrix} A^*, (\frac{\omega+2hr+2kt\pm m+1}{2},\lambda) \\ (\frac{\omega+2hr+2kt}{2},\lambda),(\frac{\omega+2hr+2kt+1}{2},\lambda),B^* \end{matrix} \right. \right] \tag{6.2.13}$$

Proof: To prove (6.2.10), let

$$f(x)=(\sin x)^{\omega-1}F_1(x)F_2(x)I(x)=\sum_{n=-\infty}^{\infty}A_n e^{-inx} \tag{6.2.14}$$

The equation (6.2.14) is valid, since $f(x)$ is continuous and bounded variation in the interval $(0,\pi)$. Multiplying both sides of (6.2.14) by e^{imx} and integrating with respect to x from 0 to π, we have

$$\int_0^\pi (\sin x)^{\omega-1} e^{imx} F_1(x)F_2(x)I(x)dx = \sum_{n=-\infty}^{\infty} A_n \int_0^\pi e^{i(m-n)x}dx$$

Now using (6.2.8) and (6.2.7), we get

$$\omega = \frac{1}{\sqrt{\pi}} e^{im\pi/2} \sum_{r,t=0}^{\infty} \phi(r)\psi(t)I_2(m) \tag{6.2.15}$$

Where $I_2(m)$ is given by (6.2.13). From (6.2.14) and (6.2.15) the Fourier exponential Series (6.2.10) is obtained.

To establish (6.2.11), let

$$(\sin x)^{\omega-1} F_1(x)F_2(x)I(x) = \frac{B_0}{2} + \sum -n = 1^\infty B_1 \cos nx \tag{6.2.16}$$

Multiplying both sides of (6.2.16) by e^{imx} and integrating with respect to x from 0 to π and using (6.2.8) and (6.2.6) we get

$$B_m = \frac{2}{\sqrt{\pi}} e^{im\pi/2} \sum_{r,t=0}^{\infty} \phi(r)\psi(t)I_2(m) \tag{6.2.17}$$

From (6.2.16) and (6.2.17), the Fourier cosine Series follows.

To prove (6.2.12), let

$$(\sin x)^{\omega-1} F_1(x)F_2(x)I(x) = \sum_{n=1}^{\infty} C_n \sin nx \tag{6.2.18}$$

Multiplying both sides of (6.2.8) and (6.2.7),we have

$$C_m = \frac{2}{\sqrt{\pi}} e^{im\pi/2} \sum_{r,t=0}^{\infty} \phi(r)\psi(t)I_2(m) \tag{6.2.19}$$

From (6.2.18) and (6.2.19), the Fourier series is obtained.

5. Double Fourier Exponential Series:

The double Exponential Fourier Series to be establishes is

$$(\sin x)^{\omega_1-1}(\sin y)^{\omega_2-1}I(xy)=\frac{1}{\pi}\sum_{n_1=-\infty}^{\infty}\sum_{n_2=-\infty}^{\infty}e^{i(n_1+n_2)\pi/2}.$$

$$I^{u+4,v}_{p_i+4,q_i+4:r}\left[z\left|\begin{matrix}A^*,(\frac{\omega_1\pm n_1+1}{2},\lambda),(\frac{\omega_2\pm n_2+1}{2},\mu)\\(\frac{\omega_1}{2},\lambda),(\frac{\omega_1+1}{2},\lambda),(\frac{\omega_2}{2},\mu),(\frac{\omega_2+1}{2},\mu),B^*\end{matrix}\right.\right]e^{-i(n_1x+n_2y)} \tag{6.2.20}$$

Where $n_1's$ and $n_2's$ are either odd or even integers; and other conditions of validity are same as stated in (6.2.9).

Proof: To prove (6.2.20), let

$$f(x,y)=(\sin x)^{\omega_1-1}(\sin y)^{\omega_2-1}I(xy)=\sum_{n_1=-\infty}^{\infty}\sum_{n_2=-\infty}^{\infty}A_{n_1,n_2}e^{-i(n_1x+n_2y)} \tag{6.2.21}$$

The equation (6.2.21) is valid, since $f(x,y)$ is continuous and of bounded variation in open interval $(0,\pi)$.

The series (6.2.21) is example of what is called a double Fourier expontial series. Instead of discussion the theory, we show one method to calculate A_{n_1,n_2} from (6.2.21). For fixed x, we note that

$$\sum_{n_1=-\infty}^{\infty}A_{n_1,n_2}e^{-in_1x}$$

Depends only on n_2; furthermore, it must be coefficient of Fourier expontial series in y of $f(x,y)$ over $0<y<\pi$.

Multiplying both sides of (6.2.21) by e^{im_2y} and integrating with respect to y from 0 to π, we get

$$(\sin x)^{\omega_1-1}\int_0^{\pi}(\sin y)^{\omega_2-1}e^{im_2y}I(xy)dxdy=\sum_{n_1=-\infty}^{\infty}\sum_{n_2=-\infty}^{\infty}A_{n_1,n_2}e^{in_1x}\int_0^{\pi}e^{i(n_1-n_2)y}dy$$

Now using (6.2.8) and (6.2.6), we have

$$\frac{(\sin x)^{\omega_1-1}}{\sqrt{\pi}} e^{im_2\pi/2} I^{u+2,v}_{p_i+2,q_i+1:r}\left[z(\sin x)^{-2\lambda} \left| \begin{matrix} A^*,(\frac{\omega_2 \pm m_2+1}{2},\mu) \\ (\frac{\omega_2}{2},\lambda)(\frac{\omega_2+1}{2},\lambda),B^* \end{matrix} \right. \right] = \sum_{n_1=-\infty}^{\infty} A_{n_1,n_2} e^{-in_1x}$$

(6.2.22)

Multiplying both sides of (6.2.22) by e^{im_1x} and integrating with respect to x from 0 to π and using (6.2.8) and (6.2.7), we get

$$A_{m_1,m_2} = \frac{1}{\pi} e^{i(m_1+m_2)\pi/2} I^{u+4,v}_{p_i+4,q_i+4:r}\left[z \left| \begin{matrix} A^*,(\frac{\omega_1 \pm m_1+1}{2},\lambda)(\frac{\omega_2 \pm m_2+1}{2},\mu) \\ (\frac{\omega_1}{2},\lambda),(\frac{\omega_1+1}{2},\lambda)(\frac{\omega_2}{2},\mu)(\frac{\omega_2+1}{2},\mu),B^* \end{matrix} \right. \right] \quad (6.2.23)$$

From (6.2.21) and (6.2.23) the double Fourier expontial series follows.

CHAPTER 7

UNIFIED STUDY

OF

ASTROPHYSICAL THERMONUCLEAR FUNCTIONS

FOR

BOLTZMANN-GIBBS STATISTICS

AND

TSALLIS STATISTICS

AND

$\overline{H}$*-FUNCTION*

We present an analytic proof of the integrals for astrophysical thermonuclear functions which are derived on the basis of Boltzmann-Gibbs statistical mechanics. Among the four different cases of astrophysical thermonuclear functions, those with a depleted high-energy tail and a cut-off at high energies find a natural interpretation in q-statistics.

1. Introduction

One of the first applications of Gamow's theory of quantum mechanical potential barrier penetration to other than his analysis of alpha radioactivity was in the field of thermonuclear astrophysics [Critchfield (1972), Mathai and Haubold (1988)]. Atkinson and Houtermans proposed that the source of energy released by stars lay in thermonuclear reactions taking place near their centers where the motion of nuclei was supposed to be in thermal equilibrium. The state of hot stellar plasmas is such that only the lightest chemical elements could contribute because of the Coulomb repulsion between nuclei. In effect, the rate of energy production is governed by the average of the Gamow penetration factor over the Maxwell-Boltzmann velocity distribution [Fowler (1984)]. The thermonuclear reaction rate is the coefficient in the rate equation that is used to describe the change of chemical composition of hot plasmas. Similarly fundamental, the inverse of this coefficient is the characteristic time scale for the respective thermonuclear reaction. Nuclear and neutrino astrophysics are vibrant fields with many experiments collecting data continuing exploring energy production in stars, particularly the Sun, deduced from neutrino measurements [Davis (2003)]. Commonly, hot stellar fusion plasmas are described in terms of Boltzmann-Gibbs statistical mechanics based on the entropy $S_{BG} = -k \sum_{i=1}^{W} p_i \ln p_i$. Thermo dynamical and Statistical description of no extensive system systems may require a generalization of Boltzmann-Gibbs approach seems to be inadequate is large self-gravitating systems and hot and turbulent plasmas.

An ultimate generalization of Boltzmann-Gibbs thermo statistics is due to Tsallis (2003) based on the following expression for entropy

$$S_q = \frac{1-\sum_{i=1}^{W} p_i^q}{q-1}, q \in R, S_1 = S_{BG}, \sum_{i=1}^{W} p_i = 1 \tag{7.1}$$

With the following expression for equal probabilities

$$S_q(p_i = 1/W, \forall i) = k\frac{W^{1-q}-1}{1-q} \equiv k\ \ln_q W, \tag{7.2}$$

$$S_1(p_i = 1/W, \forall i) = k\ \ln W = S_{BG}$$

The optimization of the entropic from (7.1) under the restriction

$$< H >_q \equiv \frac{\sum_{i=1}^{W} p_i^q E_i}{\sum_{i=1}^{W} p_j^q} = U_q, \tag{7.3}$$

Where $<\ldots>_1 = <\ldots> p_i^q / \sum_{j=1}^{W} p_j^q$ is the escort distribution, $\{E_i\}$ are the eigenvalues of the Hamiltonian H with appropriate boundary conditions, and the internal energy U is a finite fixed value, yields

$$p_i = \frac{e_q^{-\beta_q(E_i - U_q)}}{\overline{Z}_q} \tag{7.4}$$

With

$$\overline{Z}_q \equiv \sum_{j=1}^{W} e_q^{-\beta_q(E_j - U_q)}, \beta_q \equiv \frac{\beta}{\sum_{j=1}^{W} p_j^q},$$

$\beta = 1/kT$. Tsallis (2003) verifies that $q = 1$ recovers Boltzmann-Gibbs weight, $q > 1$ implies a power-law tail at high values of E_i , and $q < 1$ implies a cut-off at high values of E_i . Subsequently, this behavior is reflected in the equilibrium distribution of energies. In the following, a method for obtaining closed form representations of thermonuclear reaction rates (astrophysical thermonuclear functions) is developed, taking into account the above described behavior of q-statistics, these are the cases $q > 1$ (dubbed "depleted") and $q < 1$ (dubbed "cut-off") for the energy distribution. Thus, q-statistics provides a natural physical interpretation for the hot plasmas where deviations from the Maxwell-Boltzmann distribution are expected.

2. Definition of Astrophysical Thermonuclear Functions $\overline{I}^{d}(\nu-1,a,z,\rho)$

By employing a statistical technique, Houbold and Mathai (2000) have established the following integral formula for the derivation of closed-form representations for four astrophysical thermonuclear functions:

$$\int_0^d y^{\nu-1}\exp\left[-ay-zy^{-\rho}\right]dy \overset{def}{=} \overline{I}^{d}(\nu-1,a,z,\rho) \tag{7.5}$$

$$=\frac{d^{\nu}}{\rho}\sum_{r=0}^{\infty}\frac{(-ad)^{r}}{(r)!}\overline{H}_{1,2}^{2,0}\left[\frac{z^{1/\rho}}{d}\Big|_{(0,1/\rho),(\nu+r,1)}^{(\nu+r+1,1)}\right](\rho\neq 0), for\, d<\infty$$

$$=\frac{a^{-\nu}}{\rho}\overline{H}_{0,2}^{2,0}\left[az^{1/\rho}\Big|_{(0,1/\rho),(\nu,1)}^{-}\right](\rho\neq 0), for\, d<\infty,$$

Where $\operatorname{Re}(\nu)>0, \operatorname{Re}(a)>0, \operatorname{Re}(z)>0, \operatorname{Re}(\rho)>0$. It has shown that by the application of the Mellin-Barnes integral representationof the exponential function, the given integral can be transformed into a Mellin-Barnes type integral representing an $\overline{H}$-function [Buchaman and Srivastava (1990)]. The author also deduces the value of this integral when $\rho<0$. Results for all the four astrophysical thermonuclear functions are also obtained. The integral $\overline{I}_3(z,t,\nu)$ [Anderson, Haubold and Mathai (1994), see also Aslam Choudhary and Zubair (2002), Mathai and Haubold (1988)] is evaluated in a generalized form. For the sake of simplicity and continuation we adopt the same notations as used in [Haubold and Mathai (2000), see also Aslam Choudhary and Zubair (2002), Mathai and Haubold (1988)]. In order to provide the integral (7.5), we employ the following Mellin-Barnes integral representation for the exponential function, namely [Mathai (1993), Mathai and Saxena (1973)]

$$\exp[-x]=\frac{1}{2\pi\omega}\int_L\Gamma(s)x^{-s}ds, |x|<\infty \tag{7.6}$$

Where L is a suitable contour$\left(\omega=(-1)^{1/2}\right)$, interchange the order of integration, the given integral, denoted by $\overline{I}^{d}(\nu-1,a,z,\rho)$, transforms into the form

$$\overline{I}^{d}(\nu-1,a,z,\rho)=\frac{1}{2\pi\omega}\int_{L}\Gamma(s)z^{-s}\int_{0}^{d}y^{\nu+\rho s-1}\exp(-ay)dyds \tag{7.7}$$

We know that

$$\int_{0}^{d}\exp(-ay)y^{\nu-1}dy=d^{\nu}\sum_{r=0}^{\infty}\frac{(-ad)^{r}\Gamma(\nu+r)}{\Gamma(\nu+r+1)(r)!}for\ d<\infty,\mathrm{Re}(\nu)>0$$

$$=\Gamma(\nu)/a^{\nu},for\ d=\infty\ \mathrm{Re}(a)>0,\mathrm{Re}(\nu)>0. \tag{7.8}$$

By virtue of the above values of the y-integral for $d<\infty$ and $d=\infty$, it is found that

$$\overline{I}^{d}(\nu-1,a,z,\rho)=\frac{d^{\nu}}{\rho}\sum_{r=0}^{\infty}\frac{(-ad)^{r}}{(r)!}\int_{L}\frac{\Gamma(s/\rho)\Gamma(\nu+s+r)}{\Gamma(\nu+s+r+1)}z^{-s/\rho}ds,\quad(\rho\neq0)\,for\ d<\infty$$

(7.9)

$$=\frac{a^{-\nu}}{2\pi\omega\rho}\int_{L}\Gamma\left(\frac{s}{\rho}\right)\Gamma(\nu+s)a^{-s}z^{-s/\rho}ds\ (\rho\neq0)\,for\ d=\infty$$

The integral formula (7.5) now readily follows from the equations (7.9), if we interpret the above contour integrals in terms of the $\overline{H}$ -function, which is defined by means of a Mellin-Barnes type integral in the following manner [Buchaman and Srivastava (1990)].

$$\overline{H}_{p,q}^{m,n}[z]=\overline{H}_{p,q}^{m,n}\left[z\left|\begin{matrix}(a_j;\alpha_j;A_j)_{1,n},(a_j;\alpha_j)_{n+1,p}\\(b_j,\beta_j)_{1,m},(b_j,\beta_j;B_j)_{m+1,q}\end{matrix}\right.\right]=\frac{1}{2\pi i}\int_{-i\infty}^{i\infty}\overline{\phi}(\xi)z^{\xi}d\xi \tag{7.10}$$

Where
$$\overline{\phi}(\xi)=\frac{\prod_{j=1}^{m}\Gamma(b_j-\beta_j\xi)\prod_{j=1}^{n}\left\{\Gamma(1-a_j+\alpha_j\xi)\right\}^{A_j}}{\prod_{j=m+1}^{q}\left\{\Gamma(1-b_j+\beta_j\xi)\right\}^{B_j}\prod_{j=n+1}^{p}\Gamma(a_j-\alpha_j\xi)} \tag{7.11}$$

Here $(a_j)_{1,p}$ and $(b_j)_{1,q}$ are complex parameters, $(\alpha_j)_{1,p}, (\beta_j)_{1,q}, (A_j)_{1,n}, (B_j)_{m+1,q}$ are non-negative real numbers. $(A_j)_{1,n}, (B_j)_{m+1,q}$ can take non integer values.

A detailed and comprehensive account of the $\overline{H}$-function is available from the paper by Buschman and Srivastava (1990).In case we make the transformation $\rho = -\eta$ with $\eta > 0$ in (7.9), then the following result holds.

$$\overline{I}^d(\nu-1, a, z, -\eta) = -\frac{d^\nu}{\eta}\sum_{r=0}^{\infty}\frac{(-ad)^r}{(r)!}\overline{H}_{1,2}^{1,1}\left[z^{1/\eta}\left|\begin{matrix}(1-\nu-r+1,1;1)\\(0,1/\eta),(-\nu-r,1)\end{matrix}\right.\right] \quad \eta \neq 0, for\ z < \infty$$

(7.12)

$$= -\frac{a^{-\nu}}{\eta}\overline{H}_{1,1}^{1,1}\left[\frac{z^{1/\eta}}{a}\left|\begin{matrix}(1-\nu,1;1)\\(0,1/\eta)\end{matrix}\right.\right] \qquad \eta \neq 0, for\ z = \infty$$

(7.13)

Where $\mathrm{Re}(\nu) > 0, \mathrm{Re}(a) > 0, \mathrm{Re}(z) > 0, \eta > 0$. When ρ is real and rational then the $\overline{H}$-functions appearing in (7.5), (7.12) and (7.13) can be reduced to $\overline{G}$-functions by the application of the well known multiplication formula for Gamma functions [Erdelyi et. al (1953)]:

$$\Gamma(mz) = (2\pi)^{\frac{1-m}{2}} m^{mz-\frac{1}{2}}\Gamma(z)\Gamma\left(z+\frac{1}{m}\right)...\Gamma\left(z+\frac{m-1}{m}\right), m = 1, 2, ... \qquad (7.14)$$

3. The Astrophysical Thermonuclear Functions $\overline{I}_1(z,\nu)$ and $\overline{I}_2(z,d,\nu)$

As an application of the results (7.5), (7.12) and (7.13) we deduce the values of the following two thermonuclear functions $\overline{I}_1[]$ and $\overline{I}_2$["cut-off",Anderson, Haubold and Mathai (1994), Tsallis(2003)], in terms of which the astrophysical thermonuclear functions are expressed [Anderson, Haubold and Mathai (1994), (1988),Mathai and Saxena(1973)]:

$$\overline{I}_1(z,\nu) \stackrel{def}{=} \int_0^\infty y^{\nu-1}\exp\left[-y-\frac{z}{y^{1/2}}\right]dy = \pi^{-1/2}\overline{G}_{0,3}^{3,0}\left[\frac{z^2}{4}\left|0,\frac{1}{2},\nu\right.\right], \qquad (7.15)$$

Where $\mathrm{Re}(\nu) \geq 0$, $\mathrm{Re}(z) > 0$ and $\overline{G}_{0,3}^{3,0}(.)$ is the Meijer's generalized G-function [Mathai and Saxena (1973)].

$$\overline{I}_2(z,d,\nu) \overset{def}{=} \int_0^d y^{\nu-1} \exp\left[-y - zy^{-1/2}\right] dy = \frac{d^\nu}{\pi^{1/2}} \sum_{r=0}^{\infty} \frac{(-d)^r}{(r)!} G_{1,3}^{3,0}\left[\frac{z^2}{4d}\Big|_{(\nu+r),0,1/2}^{(\nu+r+1)}\right] \tag{7.16}$$

Where $\mathrm{Re}(\nu) \geq 1$, $\mathrm{Re}(z) > 0, d > 0$.

4. The Astrophysical Thermonuclear Functions $\overline{I}_3(z,t,\nu,\mu)$ and $\overline{I}_4(z,\delta,b,\nu)$

We now show that [Anderson, Haubold and Mathai (1994)]

$$\overline{I}_3(z,t,\nu,\mu) \overset{def}{=} \int_0^\infty y^{\nu-1} \exp\left[-\left\{y + z(y+t)^{-\mu}\right\}\right] dy \tag{7.17}$$

$$= t^\nu \sum_{r=0}^{\infty} \frac{\Gamma(\nu+r)}{(r)!} t^r \overline{H}_{2,3}^{2,1}\left[zt^{-\mu}\Big|_{(0,1),(1+\nu,\mu),(1,\mu;1)}^{(1+\nu,\mu;1),(\nu+r+1,\mu)}\right] + \sum_{r=0}^{\infty} \frac{t^r}{(r)!} \overline{H}_{2,4}^{2,2}\left[z\Big|_{(0,1),(\nu,\mu),(1,\mu;1),(\nu-r,\mu;1)}^{(1-r,\mu;1),(\nu,\mu;1)}\right],$$

Where $\mathrm{Re}(\nu) > 0$, $\mathrm{Re}(z) > 0$ and $\mu > 0$. To prove (7.17) we observe that in view of the formula (7.6), the value of the integral is equal to

$$\frac{1}{2\pi\omega} \int_L \Gamma(s) z^{-s} \int_0^\infty y^{\nu-1} [\exp(-y)] (y+t)^{s\mu} dy ds \tag{7.18}$$

Evaluating the y-integral in terms of the Whittaker function, defined in [Erdelyi et.al (1953), p.255] and [Erdelyi et.al (1953), p.257]:

$$\psi(a,c;z) = \frac{1}{\Gamma(a)} \int_0^\infty x^{a-1} e^{-zx} (1+x)^{c-a-1} dx, \ \mathrm{Re}(a) > 0, \ \mathrm{Re}(z) > 0 \tag{7.19}$$

$$= \frac{\Gamma(1-c)}{\Gamma(a-c+1)} \phi(a,c;z) + \frac{\Gamma(c-1)}{\Gamma(a)} z^{1-c} \phi(a-c+1, 2-c; z)$$

$$= \frac{\Gamma(1-c)}{\Gamma(a-c+1)} \sum_{r=0}^{\infty} \frac{(a)_r z^r}{(c)_r (r)!} + \frac{\Gamma(c-1)}{\Gamma(a)} z^{1-c} \sum_{r=0}^{\infty} \frac{(a-c+1)_r z^r}{(2-c)_r (r)!}, \tag{7.20}$$

Where c is not an integer and $\phi(a,c;z)$ is kummer's confluent hypergeometric function [Erdelyi et.al (1953), p.248], the integral (7.18) reduces to an elegant formula

$$\frac{\Gamma(\nu)t^{\nu}}{2\pi\omega}\int_{L}\Gamma(s)\psi(\nu,\nu+s\mu+1;t)z^{-s}t^{s\mu}ds \tag{7.21}$$

The integral (7.20) can be evaluated by expressing the Whittaker function appearing in its integrand in terms of its equivalent series given by (7.21) above, and reversing the order of integration and summation. Thus the given integral finally transformed into the form:

$$\overline{I}_3(z,t,\nu,\mu)=t^{\nu}\sum_{r=0}^{\infty}\frac{\Gamma(\nu+r)}{(r)!}\frac{t^{r}}{2\pi\omega}\int_{L}\frac{\Gamma(s)\Gamma(-\nu-s\mu)\Gamma(\nu+s\mu+1)}{\Gamma(-s\mu)\Gamma(\nu+r+s\mu+1)}z^{-s}t^{s\mu}ds$$

$$+\sum_{r=0}^{\infty}\frac{t^{r}}{(r)!}\frac{1}{2\pi\omega}\int_{L}\frac{\Gamma(s)\Gamma(\nu+s\mu)\Gamma(-\nu-s\mu+1)}{\Gamma(-s\mu)\Gamma(-\nu+r-s\mu+1)}z^{-s}ds \tag{7.22}$$

Which on interpreting with the help of the definition of the $\overline{H}$-function (7.10) establishes the desired result (7.17). If we set $\mu=1/2$ in (7.17), then by an appeal to Sthe duplication formula for the Gamma function (this is equation (7.14) form=2), the $\overline{H}$-function reduce to $\overline{G}$-functions [Mathai and Saxena (1973)] and consequently, we obtain

$$\overline{I}_3(z,t,\nu)\overset{def}{=}\frac{t^{\nu}}{\pi^{1/2}}\sum_{r=0}^{\infty}\frac{\Gamma(\nu+r)}{(r)!}t^{r}\overline{G}_{2,4}^{3,1}\left[\frac{z^{2}}{4t}\middle|\begin{matrix}(1+\nu;1),(\nu+r+1)\\0,1/2,(1+\nu;1),(1;1)\end{matrix}\right]$$

$$+\frac{1}{\pi^{1/2}}\sum_{r=0}^{\infty}\frac{t^{r}}{(r)!}\overline{G}_{2,5}^{3,2}\left[\frac{z^{2}}{4}\middle|\begin{matrix}(1-r;1),(\nu;1)\\0,1/2,\nu,(1;1),(\nu-r;1)\end{matrix}\right] \tag{7.23}$$

Where $\mathrm{Re}(\nu)>0,\mathrm{Re}(z)>0$. Next we will prove the formula [Anderson, Haubold and Mathai (1994)]

$$\overline{I}_4(z,\delta,b,\nu)\overset{def}{=}\int_{0}^{\infty}y^{\nu-1}\exp\left[-\left\{y+by^{\delta}+zy^{1/2}\right\}\right]dy \tag{7.24}$$

$$= \sum_{r=0}^{\infty} \frac{\left(-b/z\right)^r}{(r)!} \overline{H}_{0,2}^{2,0}\left[{}_{(r,1)(\nu+r\delta+r/2,1/2)}^{-} \right] \tag{7.25}$$

Where $\mathrm{Re}(\nu) > 0, \mathrm{Re}(z) > 0, \mathrm{Re}(b) > 0$ and $\delta > 0$. To establish the integral formula (7.25), we see that in view of (7.6), it can be written as

$$\overline{I}_4(z,\delta,b,\nu) = \int_0^{\infty} y^{\nu-1} \exp(-y) \frac{1}{2\pi\omega} \int_L \Gamma(s) y^{s/2} \left(by^{\delta+1/2} + z\right)^{-s} ds dy \tag{7.26}$$

On employing the formula

$$(1+x)^{-\alpha} = \sum_{r=0}^{\infty} \frac{(\alpha)_r}{(r)!} (-x)^r, \ |x| < 1, \tag{7.27}$$

And reversing the order of integration and summation, the equation (7.26) transforms into the form

$$\overline{I}_4(z,\delta,b,\nu) = \sum_{r=0}^{\infty} \frac{\left(-b/z\right)^r}{(r)!} \frac{1}{2\pi\omega} \int_L \Gamma(s+r) z^{-s} \int_0^{\infty} y^{\nu+\left(\delta+\frac{1}{2}\right)r+\frac{s}{2}-1} e^{-y} dy ds$$

$$= \sum_{r=0}^{\infty} \frac{\left(-b/z\right)^r}{(r)!} \frac{1}{2\pi\omega} \int_L \Gamma(s+r) \Gamma\left(\nu + \left(\delta + \frac{1}{2}\right) r + \frac{s}{2}\right) z^{-s} ds ,$$

Which, when interpreted with the help of (7.10), yields the desired result (7.25). It may be noted that the result (7.25) can be expressed in terms of the $\overline{G}$-function in the form:

$$\overline{I}_4(z,\delta,b,\nu) = \frac{1}{\pi^{1/2}} \sum_{r=0}^{\infty} \frac{(-2b/z)^r}{(r)!} \overline{G}_{0,3}^{3,0}\left[\frac{z^2}{4} \middle| {}_{\frac{r}{2}, \frac{r+1}{2}, \left(\nu+r\delta+\frac{r}{2}\right)}^{-} \right], \tag{7.28}$$

Where $\mathrm{Re}(\nu) > 0, \mathrm{Re}(z) > 0, \mathrm{Re}(b) > 0$ and $\delta > 0$. Further it is interesting to observe that, as b tends to zero, (7.28) reduces to (7.16).

5. Relation of Astrophysical Thermonuclear Functions to Kraetzel Functions

It is not out of place to mention that the function represented by the integral (7.5) in case $z=\infty$ has been studied by Kraetzel and is known as Kratzel function in the literature. It may be noted that Kraetzel (1979) introduced the integral transform

$$\left(K_{\nu}^{\rho}f\right)(x)=\int_{0}^{\infty}Z_{\rho}^{\nu}(xt)f(t)dt \quad (x>0) \tag{7.29}$$

Involving the kernel function

$$Z_{\rho}^{\nu}(x)=\int_{0}^{\infty}t^{\nu-1}\exp\left[-t^{\rho}-\frac{x}{t}\right]dt \quad (\rho>0,\nu\in C). \tag{7.30}$$

Further he has also investigated the asymptotic behavior of $Z_{\rho}^{\nu}(x)$ as $x\to 0$ and $x\to\infty$ in Kraetzel (1979)].

Finally it may be mentioned here that the integral described here for $z=\infty$ plays not only an important role in the study of astrophysical thermonuclear functions but also for Bessel-type fractional ordinary and partial differential equations [Rodrigues et al (1989)].

BIBLIOGRAPHY

Abiodum, R.F.A. and sharma, B. L. (1971).summation of series involving generalized hypergeometric functions of two variables Glasnik Math. Ser.III 6 (26),253-264.

Agal, S. N. and Koul, C.L. (1981).On integrals involving the H -function of several variables.Jnanbha 11.

Agarwal, N. (1960). A q-analogue of MacRobert's generalized E-function, Ganita, 11:49-63.

Agrawal, R.K. (1980).A multiple integrals involving the H- function of two variables. Pure Appl. Math.Sci.12, 121-126.

Agrawal, R.K. (1980a).A Study of two-dimensional integral transforms and special sfunction involving one or more variables. Ph.D. Thesis, univ. of Rajasthan, India.

Agrawal, R.P. (1965).An extension of Meijer's G-function. Proc. Nat. Inst. Sci. India Part A, 31,536-546.

Agrawal, R.P. (1970).On certain transformation formulae and Meijer's G-function of two variables. Indian J Pure Appl.Math. 1, 537-551.

Agrawal, R.P. (1988).Glimpses of the theory of generalized hypergeometric functions. Ganita Sandesh Vol.2, No. 2, 87-93.

Anandani, P. (1969). Some integral involving generalized associated Legendre's function and the H-function .Proc. Nat. Acad. Sci. India Sect. A 39, 127-136.

Anandani, P. (1969a). On Some integrals involving generalized Legendre's associated function and the H-function .Proc. Nat. Acad. Sci. India Sect. A 39, 341-348.

Anandani, P. (1969b). Some integrals involving Product of generalised Legendre associated functions and the H-function. J. Sci. Engg. Res.13, 274-279.

Anandani, P. (1969c). Some expansion formulae for H-function 1V. Rend. Circ. Math. Palerma (2) 18, 197-214.

Anandani, P. (1970). Some expansion formula for the H-function involving associated Legendre's function. J. Natu. Sci.and Math. 10, 49-51.

Anderson, W.J., Haubold, H.J., Mathai, A.M. (1994). Astrophysical Thermonuclear Functions, Astrophys. Space Sci. 214, 49.

Arora, A.K., Raina, R.K. and Koul, C.L. (1985), On the two-dimensional Weyl fractional calculus associated with the Laplace transforms, C.R. Acad. Bulg. Sci., 38(2),179-182.

Aslam Chaudhry,M., Zubair , S.M.(2002). On a class of Incomplete Gamma Functions with Applications, Chapman and Hall/CRC, Boca Raton.

Bajpai, S.D. (1968). Some expansions formulae for G-function involving Bessel functions. Proc. Indian Acad. Sci68 A, 285-290.

Bajpai, S.D. (1969). An expansion formula for fox's H-function. Proc. Cambridge philos. Soc. 65, 683-685.

Bajpai, S.D. (1969a). An expansion formula for H-function Involving Bessel functions. Labdev J. Sci. Tech. Part A 7, 18-20.

Bajpai, S.D. (1971). Some results involving Fox's H-function. Portugal Math. 30, 45-52.

Bajpai, S.D. (1980). An exponential Fourier for Fox's H-function. Math. Education 14, Sect. A, 32-34.

Bajpai, S.D. (1991). Fourier series and expansions for Fox's H-function of two variables and two-dimensional Heat equation. Hadronic Journal Val. 14, 195-201.

Bajpai, S.D. (1991a). Double and multiple half-range Fourier Series of Meijer's G-function. Acta Mathematica Vietnamica, Vol. 16, No. 1, 27-37.

Bajpai, S.D. (1991b). A new class of two-dimensional Expansion of Meijer's G-function involving Bessel Functions and Jacobi polynomials. Vijnana Parishad Anusandhan patrika, Vol. 34, No. 4, 233-236.

Bajpai, S.D. (1992). Atypical two-dimensional exponential Bessel expansion of Fox's H-function. The mathematics Education,vole. XXVI, NO. 1, 1-3.

Bajpai, S.D. And AL-Hawaj, A. Y.(1989). Meijer's G-function and ring-shaped heat conductor. Ganit (J. Bangladesh Meth. Soc.) Vol. 9,No.1, 33-38.

Bajpai, S. D. And AL-Hawaj, A.Y. (1989A). Special functions and time-domain synthesis problem .The J. of the Indian Academy of Mathematics, Vol. 11, No. 1.

Banerji, P.K. and Sethi, P.L. (1978). Operators of a generalized function of n- variables. The Mathematics Education, Vol. 16, No2, 152-159.

Barnes, E.W. (1908). A new development of the theory of the Hypergeometric functions. Proc. London Math. Soc. (2) 6, 141-177.

Basister, A.W. (1967). Transcental functions (Satisfying Non-homogeneous Linear differential equation). Macmillan, New York.

Bhise, V.M. (1962). On the derivative of Meijer's G-function And Mac Robert's e-function. Proc. Nat. Acad. Sci. India, 32, 349-354.

Bhise,V.M. (1963). Finite and infite series of Meijer's G-function and the multiplication formula for G -function. Jour. Indian Math. Soc. (N.S.) 27, 9-14.

Bhise, V.M. (1964). Some finite and infinite series of Meijer's Laplace transform. Math. Ann. 154, 267-272.

Bora, S.L. and Kalla, S.L. (1970). Some results involving Generalized function of two variables. Jyungpook Math. J. 10, 133-140.

Bora, S.L. and Kalla, S.L. (1971). An expansion formula for the generalized function of two variables. Univ. Nac. Tucuman Rev. Ser. A21, 53-58.

Bora, S.L. and Saxena, R.K. (1971).Integrals involving Product of Bessel functions and generalized hyperGeometric functions. Publ. Inst. Math. (Beograd) (N.S.) 11 (25), 23-28.

Bora, S.L. Kalla, S.L. and Saxena, R.K. (1970). On integral Transforms. Univ. Mac. Tucuman Rev. Ser. A 20, 181-188.

Bora, S.L., Saxena, R.K. and Kalla, S.L. (1972). An expansion Formula for fox's H -function of two variables. Univ. Nac. Tucuman Rev. Ser. A 22, 43-48.

Braaksma, B.L.J. (1963). Asymptotic expansions and analytic Continuations for a class of Barnes-integrals. Composito Math. 15, 339-341.

Bromwich, T.J.I' A. (1931). An introduction to the theory of infinite series, 2nd ed. Macmillan and Co., London.

Buchman, R.G. (1972). Contiguous relations are related Formulas for the H-function of Fox. Jnanabha Sect. A 2, 39-47.

Buchman, R.G. (1979). Integral equations and Laplace transformations. J. Indian Acad. Math. 1, No. 1, 1-4.

Buschman, R.G. and Srivastava, H.M. (1990). The $\overline{H}$ – function associated with a certain class of Feynman integrals, J. Phys. A: Math. Gen. 23, 4,707-710.

Buschman, R.G., Koul, C.L. and Gupta, K.C. (1977). Convolution integral equations involving the function askernel, Glasnik Mat. Ser. III 12 (32), 61-66.

Chaturvedi, K.K.and Royal, A.N. (1972). A*-function. Indian J. Pure Appl. Math. 3, 357-360.

Chaurasia, V.B.L. (1988). Integration of certain products Associated with the multivariable H -function. Gahnita Sandesh Vol. 2, No. 2, 66-69.

Chaurasia, V.B.L. (1991). Integrals involving generalized Lauricella's function and the H-function of several Complex variables. Indian J. Pure Appl. Math. 22(5). 397-401.

Chaurasia,V.B.L and Agnihotri,A.(2010).Two dimensional generalized weyl fractional calculus and special functions. Tamkang journal of Mathematics, volume 41,No. 2,pp 139-148.

Chaurasia,V.B.L and Srivastava,A.(2006).Two dimensional generalized weyl fractional calculus pertaining to two dimensional $\overline{H}$ -transforms. Tamkang journal of Mathematics, volume 37,No. 3,pp 237-249.

Chaurasia, V.B.L. and Tyagi, Sanjeev (1991). Integrals Involving generalized Lauricella's function and the H-function of several complex variables. Indian J. Pure Appl. Math. 22(5), 397-401.

Churchill, R.V. (1961). Fourier series and boundary value Problems. Mc-Grow Hill, New York.

Critchfield,C.L. (1972). in:Comology Fusion and Other Matters: George Gamow Memorial Volume, Ed. F. Reines, Colorado Associated University Press, Colorado, 186.

Davis Jr., R. (2003). Nuclear and neutrino astrophysics, Rev. Mod. Phys. 75, 985.

Detach, Gustav (1950). Handbuch der Laplace-transformation. Vol. 1, Verlag Birkhauser, Basel.

Dighe, M. And Baishe, V.M. (1979/80). On composition of Fractional integral operators as an integral operator with flourier type kernel. Math. Notate 27, 237.

Ditkin, V.A. and Prudnikov, A.P. (1962). Operational Calculus in Two Variables and its Applications, New York.

Ditkin, V.A. and prudnikov, A.P. (1962). Operational Calculus in two variables and its applications. Translated from the Russian by D.M.G. Wishart. International series of monographs on pure and Appl. Math. Vol; 24, Pergamon press, Oxford.

Dixit, A.K. (1981). On some integral relations involving the general H-function of several complex variables. Indian J. pure Appl. Math. 12, 977-983.

Doetseh, Gustav (1974). Introduction to the theory and applications of the Laplace transformation. Springer verlag, Berlin Heidelberg, New York.

Erdelyi, A. (1940). On fractional integration and its application of the theory of Hankel transform. Quart. J.Math. (Oxford) (1) 11, 293-303.

Erdelyi, A. (1950-51). On some functional transformations.Univ. Politec. Torino Rend. Sem. Mat. 10, 217-234.

Erdelyi, A. (1951). On a generalization of the Laplace transformation. Proc. Edinburgh Math. Soc. (2) 10, 53-55.

Erdelyi, A. (1975). Fractional integrals of generalized Functions, Fractional calculus and its applications. (B. Ross, ed.). Springer-verlag, Berlin, Heidelberg and NewYork.

Erdelyi, A. (1954).Higher Transcendental Functions, vol. I, II, McGraw-Hill, New York.

Erdelyi,A.et.al. (1953). Higher Transcendental Functions ,Vol.1,Mc.Graw-Hill, New York.

Erdelyi,A. et. Al. (1954). Tables of Integral Transforms, Vol. II, McGraw-Hill, New-York.

Erdelyi, A. And kober, h. (1940). Some remarks on Hankel Transforms. Quart. J. Math. (Oxford) 11, 212-221.

Erdelyi, A., Magnus, W., Oberhettinger, F. And Tricomi, F.g. (1953). Higher transcendenatal functions, Vol. I and II, McGraw-Hill, New York.

Erdelyi, A., Magnus, W., Oberhettinger, F. And Tricomi, F.g. (1954). Tables of integral transforms, Vols. I and II. MaGraw-Hill, New York.

Erdelyi, A., Magnus, W., Oberhettinger, F. And Tricomi, F.G. (1955). Higher transcendental functions, Vol. III. McGraw-Hill, New York.

Exton, Harold (1976). Multiple hypergeometric functions and applications. Ellis Horwood Ltd., Chichester.

Exton, Harold (1978). Handbook of hypergeometric integrals. Ellis Horwood Ltd., Chichester.

Fowler , W.A. (1984). Maxwell-Boltzmann velocity distribution, Rev. Mod. Phys., 56, 149.

Fox, C. (1928). The asymptotic expansion of generalized Hypergeometric functions. Proc. London Math. Soc. (2) 27, 389-400.

Fox, c. (1961). The G-and H-functions as symmetrical Fourier kernels. Trans. Amer. Math. Soc. 98, 395-429 .

Fox, C. (1963). Integral transforms based upon fractional Integration. Proc. Cambridge philos. Soc. 59, 63-71.

Fox, C. (1965). A formal solution of certain dual integral Equations. Trans. Amer. Math. Soc. 119, 389-398.

Fox, C. (1971). Solving integral equations by L and L^{-1} operators. Proc. Amer. Math. Soc. 29, 299-306.

Fox, C. (1972). Application of Laplace transforms and their Inverses. Proc. Amer. Math. Soc. 35, 193-200.

Garg, M. (1980). A note on a generating function for the multivariable H-function. Jnanabha (professor Arthus Erdelyi Memorial Volume), 9/10.

Garg, R.S. (1981). Some general integral relations for the multivariable H-function. Jnanabha.

Gasper, G. and Rahman, M. (1990). Basic Hypergeometric Series, Cambridge University Press, Cambridge.

Gautam, G.P. and Goyal, A.N. (1981). On Fourier kernels and asymptotic expansion. Indian J. Pure Appl. Math. 12, 109-1105.

Gautam, G.P. and Goyal, A.N. (1983). On a multivariable A-function. Univ. Nac. Tucuman REV. Ser. A 21.

Gokhroo, D.C. (1974). Sum of an infinite series involving meijer's G-function. Proc. Nat. Acad. Sci. India44 (A) I. 65-68.

Goyal, S.P. (1970). On some finite integrals involving generalized G -function. Proc. Nat. Acad. Sci. IndiaSect. A 40, 219-228.

Goyal, S.P. and Agarwal, R.K. (1980).Basic series relations for the H-function of two variables. Pure Appl. Math. Sci. 11, 39-48.

Goyal, S.P. and Agawal, R.K. (1980a). A two-dimensional Neumann expansion for the H-function of two variables. Kyungpook Math. J. 20, 95-103.

Goyal, S.P. and Agrawal, R.K. (1980b). The distributions of a linear combination and the ratio of product of random variables associated with the multivariable H-function. Jnanabha (professor Arthur Erdelyi Memorial Volume) 9/10.

Goyal, S.P. and Agrawal, R.K. (1981). Bivariate distributions and multivariable H-function. Indian J. Pure Appl. Math.12, 380-387.

Goyal, S.P. and Jain, R.N. (1987). Fractional integral operatiors and the generalized hypergeometric functions. Indian J. Pure Appl. Math. 18, 6, 251-259.

Goyal, S.P., Jain, R.N. and Gaur, Neelima (1991). Fractional integral operators involving a product of generalized hypergeometric functions and a general class of polynomials. Indian J. Pure Appl. Math. 22(5), 403-411.

Gradshteyn, I.S. and Ryzhik, I.M. (1963). Tables of Integrals, Series and Products, Academic Press, NewYork.

Gradshteyn, I.S. and Ryzhik, I.M. (1965). Tables of integrals, Series and products. Academic press, New York.

Gradsteyn, I.S. and Ryzhik, I. M. (1980). Tables of Integrals, Series and Products. Corrected and Enlarged Eddition, Academic Press. Inc.

Gupta, K.C. (1965). On the H-function. Ann. Soc. Sci.Bruxlles Ser. I. 79, 97-106.

Gupta, K.C. (1966). Integrals involving the H-functions. Proc. Nat. Acad. Sci. India Sect. A 36, 504-509.

Gupta, K.C. and Agrawal, S.M. (1989). A finite integral involving a general class of polynomials and the multivariable H-function. Indian J. Pure Appl. Math. 20(6), 604-608.

Gupta, K.C. and Garg, Mridula (1984). A study of some multi- dimensional fractional integral operators. Jnanabha Vol. 14, 53-69.

Gupta, K.C. and Jain, U.C. (1966). The H-function-II. Proc. Nat. Acad. Sci. India Sect. A 36, 594-609.

Gupta, K.C. and Jain, U.C. (1968).On the derivative of the H-function. Proc. Nat. Acad. Sci. India Sect. A 38, 189-192.

Gupta, K.C. and Jain, U.C. (1969). The H-function - Iv. Vijnana parishad Anusandhan patrika 12, 25-30.

Gupta, K.C. and Koul, C.L. (1977): An integral involving H-function, Math. Student, 45, 33-38.

Gupta, K.C. and Mittal, P.K. (1970). The H-function transform. J. Austral. Math. Soc. 11, 142-148.

Gupta, K.C. and Mittal, P.K. (1971). The H-function transform. II. J. Austral. Math. Soc. 12, 444-450.

Gupta, K.C. and Srivastava, A.(1972). On finite expansions for the H-function. Indian J. Pure Appl. Math. 3, 322-328.

Gupta,K.C. , Jain, Rashmi and Sharma,Arti (2003). A study of unified finite integral with applications, J.Rajashthan Acad. Phys. Sci. Vol.2,No.4, 269-282.

Gupta, P.N. and Rathie, P.N. (1968). Some results involving generalized hypergeometric functions. Rev. Mat. Univ. Purma (2) 9, 91-96.

Gupta, Rajni (1988). Fractional integral operators and a general class of polynomials. J. Indian Acad. Math. Vol. 10, No. 1, 32-37.

Gupta,Rajni (1988). Transformation formulas for the multivariable H-function , Indian J. Of Mathematics, Vol. 30,N0.2,105-118.

Gupta, Rajni (1990). Fractional integraloperators and a general class of polynomials. Indian J. of Mathematics Vol. 32, No. 1, 69-77.

Gupta, R.K. ,Saxena, R. K. (1992). On certain transformations of multiple basic Hypergeometric functions J. Math. Phy. Sci., 26, no.4, 413- 418.

Hardy, G.H. and Littlewood, J.E. (1925). Some properties of fractional integrals. Proc. London Math. Soc. (2), 24, 37-41.

Hathie, N.R. (1979). Integrals involving H-function. Vijnana Parishad Anusandhan Patrika 22, 253-258.

Hathie, N.R. (1980). Integrals involving H-function. II. Vijnana Parishad Anusandhan Patrika 23, 331-337.

Haubold, H.J. and Mathai, A.M. (1984). On the nuclear energy generation rate in a simple analytic stellar mode, Annalen der Physik, 41(6), 372-379.

Haubold, H.J. and Mathai, A.M.(1984a). On nuclear reaction rate theory, Annalen der Physik, 41(6), 380-396.

Haubold, H.J. and Mathai, A.M. (1985). The Maxwell-Boltzmann approach to the nuclear reaction rate theory, Forschr. Phys., 33, 623-644.

Haubold, H.J. and Mathai, A.M. (1986). The resonant thermonuclear reaction rate, J.Math. Phys., 27,2203-2207.

Haubold, H.J. and Mathai, A.M. (1986b). Analytical solution to the problem of nuclear energy generation rate in a simple stellar model, Astron. Nachr., 307, 9-12.

Haubold, H.J. and Mathai, A.M. (1986c). Analytic results for screened non-resonant nuclear reaction rates, Astrophysics and Space Science, 127, 45-53.

Haubold , H.J. and Mathai , A.M.(2000). The fractional Kinetic equation and thermonuclear functions, Astrophysics and Space Science 273, 53-63.

Inayat-Hussian,A.A. (1987). New properties of hypergeometric series derivable from Feynman Integrals II. A generalization of the H-function. J Phy. A: Math. 20, 4119-4128.

Jain, R. (1992): A study of multidimensional Laplace transform and its applications, Soochow J.of Math., 183-193.

Jain, R.N. (1966). A finite series of G-function. Math. Japan, 11, 129-131.

Jain, R.N. (1969). General series involving H-functions. Proc. Cambridge philos. Soc. 65, 461-465.

Jain, U.C. (1967). Certain recurrence relations for the H-function. Proc. Nat. Inat. Sci. India parta 33, 19-24.

Jain, U.C. (1968). On an integral involving the H-function. J. Austral. Math. Soc. 8, 373-376.

Joshi, C.M. and Arya, J.P. (1981). Concerning some properties of the H-function of several variables. Indian J. Pure Appl. Math. 12, 826-843.

Kalla, S.L. (1966).Some theorems on fractional integration. Proc. Nat. Acad. Sci. India Sect. A 36, 1007-1012.

Kalla, S.L. (1969).Some theorems on fractional integration II. Proc. Nat. Acad. Sci. India Sect. A 39, 44-56.

Kalla, S.L. (1969). Integral operators involving Fox's H-function. Acat. Maxicana Ci. Tecn. 3, 117-122.

Ka!la, S.L. (1970). Fractional integration operators involving generalized hypergeometric functions. Univ. Nac. Tucuman Rev. Ser. A. 20, 93-100.

Kalla, S.L. (1970-71). On operators of fractional integration. Math. Notae 22, 89-93.

Kalla, S.L. (1976).On operators of fractional integration. II. Math. Notae 25, 29-35.

Kalla,S.L. and kiryakova, V.S. (1990). An H-function generalized fractional calculus based upon compositions of Erdelyi-kober operators in L_p. Math. Japan. 35, 1151-1171.

Kalla, S.L. and Munot, P.C. (1970). An expansion formula for the generalized Fox's H-function of two variables. Repub. Venezuela Bot. Acad. Ci Fis. Math. Natur. 30, 87-93.

Kalla, S.L. and Saxena, R.K. (1969). Integral operators involving hypergeometric functions. Math. Z. 108, 231-234.

Khadia, S.S. and Goyal, A.N. (1970). On the generalized function of 'n' variables. II. Vijnana parishad Anusandhan patrika 13, 191-201.

Khadia, S.S. and Goyal, A.N. (1975).On the generalized function of'n' variables. II. Vijnana parished Anusandhan patrika 18, 359-365.

Kim,Y.C. and Srivastava,H.M.(1998).Fractional integral and other linear operators associated with the Gaussian Hypergeometric function.Complex variable theory Appl.34.293-312.

Kiryakova, V.S. (1986).On operators of fractional integration involving Meijer's G -function, Compt. Acad. Bulg. Sci. 39, 25-28.

Kiryakova, V.S. (1988). Fractional integration operators involving Fox's $H_{m,m}^{m,0}$ -function. Compt. Rend. Bulg. Sci. 41, 11-14.

Kober, H. (1940). On fractional integrals and derivatives, Quart. J. Math. Oxford. Ser. 11, 193-211.

Korganoff, V. (1980). Introduction to Astrophysics, Reidel Publication, Boston.

Kraetzel, E. (1979). in: Generalized Functions and Operational Calculus, Proc. Intern. Conf. Varna, 1975, Bulg. Acad. Sci., Sofia, 148.

Kumbhat, R.K. and Saxena, R.K. (1974). A formal solution of certain triple integral equations involving H -functions, Proc. Nat. Acad. Sci. India, Sect. A 44, 153-159.

Kumbhat, R.K. and Saxena R.K. (1975). Theorems connecting L , L^{-1} and fractional integration operators. Proc. Nat. Acad. Sci. India 45(A) III, 205-209.

Lawrynowicz, J. (1969). Remarks on the preceding paper of P .Anandani. Ann. Polon. Math. 21, 120-123.

Lebedev, V.N. (1965). Special functions and their applications. (Translated from the Russian by Richard A. Silverman) Prentice-Hall, New Jersey.

Lerch, E. (1903). Sur Un. Point de la theories des functions generatrices d' Abel, Acta Math. 27, 339.

Lin, S.-D. , Liu, S.-J. and Srivastava, H. M. (2011), Some families of hypergeometric polynomials and associated multiple integral representations, Integral Transforms Spec. Funct. 22. 403-414.

Liouville, J. (1832a).Memoire Sur le calcul des differentielles a indices quelconques. J.Ecole polytech., Sect. 21, 13, 71-162.

Liouville, J. (1832). Memoire Sur quelques question de geometrie et de mecanique, et Sur un nouveau genre de calcul pour resoudre ces questions. J. Ecole polytech. , Sect. 21, 13, 1-69.

Loonker,D. and Banerjee P.K.(2004).Distributional Laplace-Hankel transform by fractional integral operators. The Mathematics student.vol 73 , No.1-4.

Love, E.R. (1967). Some integral equations involving hypergeometric functions. Edinburg Math. Soc. (3) 15, 169-198.

Love, E.R. (1970). Changing the order of integration. J. Austral. Math. Soc. 11, 421-432.

Lowndes, J.S. (1970). A generalization of the Erdelyi-kober operators. Proc. Edinburgh Math. Soc. Ser. II. 17, 139-148.

Lowndes, J.S. (1985). On some fractional integrals and their applications. Proc. Of the Edinburg Math. Soc. 28, 97-105.

Luke, Y.L. (1969). The special functions and their approximations, Vols. I and II. Academic press, New York.

Luke, Y.L. (1975). Mathematical functions and their approximation. Academic press, New York.

MacRobert, T.M. (1962). Functions of a complex variables. 5th ed. Macmillan, London.

Marichev, O.I. (1983). Handbook of integral transforms of higher transcendental function, theory and algorithmic tables. Ellis Harwood Ltd., Chichester.

Mathai , A.M. (1993). A Handbook of Generalized Special Functions for Statistics and Physical Sciences, Oxford University Press, Oxford.

Mathai, A.M. and Haubold , H.J. (1988).Modern Problems in Nuclear and Neutrino Astrophysics, Akademie-Verlag, Berlin.

Mathai, A.M. and Saxena, R.K. (1966). On a generalized hypergeometric distribution. Metrika 11, 127-132.

Mathai, A.M. and Saxena, R.K. (1973). Generalized Hypergeometric Functions With Applications in Statistics and Physical Sciences, Springer-Verlag, Berlin.

Mathai, A.M. and Saxena, R.K. (1973). Generalized hypergeometric functions with applications in statistics and physical Sciences, Springer-Verlag, Lecture Notes No. 348, Heidelberg and New York.

Mathai, A.M. and Saxena, R.K. (1978). The H-function With Applications in Statistics and Other Disciplines, John Wiley and Sons. Inc. New York.

Mathai, A.M. and Saxena, R.K. (1978). The H-function with applications in statistics and other Disciplines. Wiley Eastern Ltd., New Delhi.

Mathai,A.M.,Saxena,R.K.,Haubold,H.J.(2004).Astrophysical Thermonuclear functions for Boltzmann-Gibbs statistics and Tsallis statistics,Physica A 344,649-656.

Mathur, B.L. (1981).Some results concerning a special function of several complex variables. Indian J.Pure Appl. Math. 12, 1001- 1006.

Mathur, B.L. and Krishna, S. (1977). On multivariate fractional integration operators. Indian J. Pure Appl. Maths. 8, 1078-1082.

McBride, A.c. and Roach, G.F. (1985). Fractional calculus. pitman Advanced publishing program .

Meijer, C.S. (1941). Multiplikations theorems fur die Function $G_{p,q}^{m,n}(z)$. Nederl. Akad. Wetensch. Proc. 44, 727-737 =Indag. Math. 3, 338-348.

Meijer, C.S. (1946). On the G-function I-VIII, Nederl. Akad. wetenach. Proc. 49, 227-237, 344-356, 457-469, 632-641, 765-772, 936-943, 1061-1072, 1165-1175=Indag. Math. 8, 124-134, 213-225, 312-324, 391-400, 468-475, 595-602, 661-670, 713-723.

Meijer, C.S. (1952-1956). Expansion theorems for the G- function I-XI. Nederl. Akad. Wetenach. Proc. Ser. A, 55=Indag. Math. 14, 369-379, 483-487 (1952), Proc. Ser. A. 56=Indag. Math. 15, 43-49, 187-193, 349-357(1953), Proc. Ser. A. 57=Indag. Math. 16, 77-82, 83-91, 273-279 (1954), Proc. Ser. A. 58=Indag. Math. 17, 243-251, 309-314 (1955), Proc. Ser. A, 59=Indag. Math. 18, 70-82 (1956).

Meijer, C.S. (1964). On the G-function, I. Proc. Nederl. Akad. Wetensch. 49, 227-237.

Miller,A.M.and Srivastava,H.M.(1998).On the Mellin transform of a product of Hypergeometric function.J.Austral,Math.Soc.Ser.B 40.222-237.

Miller,K.S. (1975).The Weyl fractional calculus,(Lecture Notes in Math. 457), Berlin 80-89.

Milne-Thomson, L.M. (1933). The Calculus of Finite Differences. Macmillan, London.

Mishra, A.P. (1975).On a fractional differential operator, Ganita 26(2) ,1-18

Mishra, S. (1990). Integrals involving expontial function, generalized hypergeometric series and Fox's H-function and Fourier series for products of generalized hypergeometric functions,J. Indian Acad. Math.,12(1), 33-47.

Mittal, P.K. and Gupta, K.C. (1972). An integral involving generalized function of two variables. Proc. Indian Acad. Sci. Sect. A 75, 117-123.

Mourya, D.P. (1970). Fractional integrals of the function of two variables. Proc. Indian Acad. Sci. Sect. A 72, 173-184.

Munot, P.C. and Kalla, S.L. (1971). On an extension of generealized H-function of two variables. Univ. Nac. Tucuman Rev. Ser. A 21, 67-84.

Munot, P.C. and Mathur, R. (1983). Certain integral relations for the multivariable H-function. Indian J. Pure Appl. Math. 14(8), 955-964.

Nair, V.C. and Nambudripad, B.N. (1973). Integration of H-function with respect to their parameters. Proc. Nat. Acad. Sci. India, 43(A), IV.

Nair, V.C. and Samar, H.S. (1971A). The product of two H-function expressed as a finite integral of the sum of a series of H-functions. Math. Education Sect. A, 5, 45-48.

Nair, V.C. and Samar, M.S. (1971). An integral involving the product of three H-functions. Math. Nachr. 49, 101-105.

Nair, V.C. and Samar, M.S. (1975). A relation between the Laplace transform and the Mellin transform with applications, Portugaliae Mathematica, vol.34, Fasc. 3 , 149-155.

Nishimoto, K. (1984). Fractional caslculus. Vol. I, II and III. Descartes Press Co. Koriyama, Japan.

Nishimoto, K. (1991). An Essence of Nishimoto's Fractional calculus. Descartes Press Co. Koriyama, Japan.

Okokiolu, G.O. (1967). Fractional integrals of H- type,Quart. J. Math. Oxford. 18(2), 33-42.

Oldham, Keith, B. And Spemier, J. (1974). The fractional calculus. Academic press, New Yourk.

Olkha, G.S. (1970). Some finite expansions for the H-function. Indian J. Pure Appl. Math. 1, 425-429.

Panda, R. (1977). Certiain integrals involving the H -function of several variables. Publ. Inst. Math. (Beograd) (N.S.), 22 (36), 207-210.

Panda, R. (1977a). Integration of certain products associated with the H-function of several complex variables. Comment. Math. Univ. St. Paul. 26, 115-123.

Panda, R. (1977b). On a multiple integral involving the H-function of several complex variables. Indian J. Math., 19, 157-162.

Pandey, R.N. and Pandey, S.K. (1985). Power series expansions of H-function of n-variables. Vijnana parished Anusandhan patrika 28(2), 137-149.

Parashar, B.P. (1967). Fourier series for H-function. Proc. Cambridge philos. Soc. 63, 1083-1085.

Parashar, B.P. (1968). Domain and Range of fractional integration. Math. Japan 12, 141-145.

Pathak, R.S. (1970). Some results involving G- and H-functions. Bull.cal. Math. Soc. 62, 97.

Pathak, R.S. and Pandey, R.N. (1989). On certain fractional integral operators. Bull. Cal. Math. Soc. 81, 17-24.

Pathan, A.M. (1968). Certain recurrence relations. Proc. Cambridge Phills. Soc. 64, 1045-1048.

Prasad, Y.N. (1986). Multivariable I–function. Vijnana Parishad Anusandhan Patrika, 29, 232-235.

Prasad, Y.N. and Singh, S.N. (1977). Multiple integrals involving the generalized function of two variables. Ann. Fac. Sic. Univ. Nat. Zaire (Kinshasa) Sect. Math.Phys. 3, 252-265.

Ragab,F.M. (1957).Integration of certain fractional integral operators. Proc. Glasgow Math. Assoc. 3, 94-98.

Raina, R.K. (1984). On composition of certain fractional integral operators. Indian J. Pure Appl. Math. 15, 509-516.

Raina, R.K., Kiryakova,V.S. (1983). On the Weyl fractional operator of two dimensions, C.R. Acad. Bulg. Sci., 36(10),1273-1276.

Rainville, E.D. (1960). Special functions. The Macmillan co., New York.

Rainville, E.D. (1963). Special functions. The Macmillan co.Inc, New York.

Rainville, E.D. (1963).The Laplace transform: An introduction, The Macmillan Co., New York.

Ramawat ,A. ,Singh, Y. and Saxena ,R. K. (1992).Multidimensional fractional integration operators associated with multivariable I-function. Ganita Sandesh 6 , 105-112.

Rathie, A.K. (1997). A new generalization of generalized hypergeometric functions, Le Mathematiche Fasc. II, 52, 297-310.

Riemann, B. (1876). Gesammelte Werke, pp. 331-344.

Riesz , M. (1950). L-integrable de Riemann –Liouville et le probleme de Cauchy, Acta Math. Japonica, 81, 1-222.

Rodrigues, J.,Trujillo,J.J., Rivero, M. (1989). in: Differential Equations, Xanthi, 1987, Lecture Notes in Pure and Applied Mathematics, Series No. 118, Denver, New York, 613.

Ross,B. (1975) . Fractional calculus and its applications .Lecture Notes in Maths. 457, springerverlag, New York.

Saigo M. (1984). A generalization of fractional calculus, Fractional Calculus, Research Notes in Maths. 138, pitman.

Saigo, M.,Ram, J. and Saxena ,R. K. (1992). On the fractional calculus operator associated with H-function, Ganita Sandesh 636-47.

Saigo, M.,Ram J. and Saxena R. K. (1995). On the two dimensional Weyl fractional Calculus associated with two dimensional H-transform, J. Fract. Calc. 8, 1995, pp.63-73.

Samko, S., Kilbas, A. and Maricher, O.I. (1987). Integrals and derivatives of fractional order and some of their applications (In Russian). Nauka i Technika, Minsk.

Saxena, R.K. (1960). An integral involving G- function. Proc. Nat. Inst. Sci. India Part A 26, 661-664.

Saxena, R.K. (1966). An inversion formula for a kernel involving a Mellin-Barnes type integral. Amer. Math. Soc.17, 771-779.

Saxena, R.K. (1967). A formal solution of certain dual integral equations involving H-function. Proc. Cambridge Philos. Soc. 63, 171-178.

Saxena, R.K. (1967a). On the formal solution of dual integral equations. Proc. Amer. Math. Soc. 18, 1-8.

Saxena, R.K. (1967b). On fractional integration operators Math. Z., 96, 288-291.

Saxena, R.K. (1970). On the H-function of n-variables. Kyungpook Math. J. 17, 221-226.

Saxena, R.K. (1971). Integrals of products of H-functions. Univ. Nac. Tucuman Rev. Ser. A 21, 185-191.

Saxena, R.K. (1971a). Definite integrals involving Fox's H–function. Acta Maxicana Ci. Tecn. 5, 6-11.

Saxena, R.K. (1971b). An integral associated with generalized H-function and Whittaker functions. Acta Mexicana C1. Tecn. 5, 149-154.

Sxena, R.K. (1974). On a generalized function of n-variables. Kyungpook Math. J. 14, 225-259.

Saxena, R.K. and Agrawal, Suveen (1985). Certain integrals for multivariate H -function involving Jacobi polynomials and kampe de Fariet function. Vijnana parishad Anusandhan patrika, Vol. 28, No. 2, 113-127.

Saxena ,R. K. and Gupta, O.P. (1993).On certain relations connecting Erd´elyi-Kober operators and generalized Laplace transform. J. Fract. Calc. 3, pp.73-79.

Saxena, R.K. and Kumar, R. (1990). Recurrence relations for the basic analogue of the H -function, J. Nat. Acad. Math. 8: 48-54.

Saxena, R.K. and Kumbhat, R.K. (1973). A generalization of the kober operators. Vijnana parishad Anusandhan patrika 16, 31-36.

Saxena, R.K. and Kumbhat, R.K. (1973a). Fractional integral operator of two variables. Proc. Indian Acad. Sci. Sect. A 78, 177-186.

Saxena, R.K. and Kumbhat, R. K. (1974). Integral operators involving H -function. Indian J. Pure Appl. Maths. 5, 1-6.

Saxena, R.K. and Kumbhat, R.K. (1975). Some properties of generalized kober operators. Vijnana parishad Anusandhan patrika 18, 139-150.

Saxena, R.K. and Mathur, S.N. (1971). A finite series of the H -function. Univ. Nac. Tucuman Rev. Ser. A 21, 49-52.

Saxena, R.K. and Modi, G.C. (1974).Some expansions involving H -function of two variables. C.R. Acad. Bulgare Sci. 27, 165-168.

Saxena, R.K. and Modi, G.C. (1974a). Expansion formulae of the generalized H -function. Vijnana parishad Anusandhan patrika 17, 185-195.

Saxena, R.K. and Modi, G.C. (1975). Generalized H -function as a symmetrical Fourier kernel. Univ. Nac. Tucuman Rev. Ser. A, 25.

Saxena, R.K. and Modi, G.C. (1980). Multidimensional fractional integration operators associated with hypergeometric functions. Nat. Acad. Sci. Letters, Vol. 3, No. 5, 155-157.

Saxena, R.K. and Modi, G.C. (1985). Multidimensional fractional integration operators associated with hypergeometric function II. Vijnana parishad Anusandhan patrika, Vol. 28, No. 1, 87-97.

Saxena, R.K., Gupta, O.P. and Kumbhat, R.K. (1989). On the two-dimensional Weyl fractional calculus,C.R. Acad. Bulg. Sci.,42(7) 11-14.

Saxena, R.K., Modi, G.C. and Kalla, S.L. (1983). A basic analogue of Fox's H-function, Rev. Tec. Ing. Univ. Zulia, 6: 139-143.

Saxena, R. K., Ram, C.(1990). An integral involving multivariable H-function. Vijnana Parishad Anusandhan Patrika, 33, No.2, pp.87-93.

Saxena ,R. K., Ram, C. and Dave, O. P. (1994). Integrals associated with Gauss' hypergeometric series, Multivariable H-function an a general class of polynomials, Jnanabha, 14, pp35-48.

Saxena ,R. K. ,Ram, C. and Dudi ,N.(2005). Some results on a new generalized hypergeometric function, Acta Ciencia Indica, Math.Vol. 31 M 1265-1272.

Saxena, R.K., Ram, J.(1990). On the two-dimensional Whittaker transforms, SERDICA Bulgaricae Mathematicae publicationes, vol. 16, 27-30.

Saxena ,R. K., Ram, J.and Chandak ,S. (2007). Unified fractional integral formulas involving the I-function associated with the modified Saigo operator, Acta Ciencia India Math. Vol 33M, No. 3, pp 693-704.

Saxena, R. K., Ram, J., Chandak ,S.and Kalla ,S.L. (2008).Unified fractional integral formulas for the Fox-Wright generalized hypergeometric function, Kuwait J. Sci. Eng. 351-20.

Saxena ,R. K., Ram, J.and Suthar, D.L. (2005).Integral Formulas for the H-function Generalized Fractional Calculus associated with Erd´elyi-Kober operator of Weyl type, Acta Ciencia Indica, Math.Vol. XXXI M, No.3, 761-766.

Saxena, R. K., Ram, J.and Suthar, D.L. (2006).Certain properties of generalized fractional integral operators associated with Mellin and Laplace transformations, J.Indian Acad. Math. Vol.28, No.1, pp.166-177.

Saxena ,R. K. ,Singh, Y. (1991). Integral operators involving Multivariable I-function. Proc. Nat. Acad. Sci. India,61(A), no.2, pp. 177-183.

Saxena R. K.,Singh, Y. and Ramawat, A. (1993). On compositions of generalized fractional integrals. Hadronic J. Supplement 8, 137-159 .

Saxena, R.K. , Yadav, R.K., Purohit, S.D. and Kalla, S.L. (2005). Kober fractional q-integral operator of the basic analogue of the H-function, Rev. Tec. Ing. Univ. Zulia, 28(2): 154-158.

Saxena,V.P. (1982): Formal solutions of certain new pair of dual integral equations involving H-functions, Proc.Nat.Acad.Sci. India A (52), 366-375.

Shah, M. (1969). Some results on Fourier series for H-functions. J. Natur. Sci. and Math. 9, 121-131.

Shah, M. (1973). Application of Hermite polynomials for certain properties of fox's H-function of two variables. Univ. Nac. Tucuman Rev. Ser. A 23, 165-178.

Sharma, B.L. (1965). On a generalized function of two variables. I. Ann. Soc. Bruxeles Ser. I, 79, 26-40.

Sharma, B.L. (1968). A new expansion formula for hypergeometric functions of two variables. Proc. Cambridge philos. Soc. 64, 413-416.

Sharma ,K.C.and Singh, Roshni.(2010).Finite summation formulae for multivariable H-function. Tamkang journal of Mathematics, volume 41,No. 4,pp 375-377.

Sharma, O.P. (1965). Some finite and infinite integrals involving H-function and Gauss's hypergeometric functions. Collect. Math.17,197-209.

Siddiqui, A. (1979). Integral involving the H-function of several variables and an integral function of two complex variables. Bull. Inst. Math. Acad. Sinica 7, 329-332.

Singh, Ratan (1968). On some results involving H-function of fox. Proc. Nat. Acad. Sci. India Sect. A 38, 240- 322.

Singh, Ratan (1970). An inversion formula for fox H-transform. Proc. Nat. Acad. Sci. India Sect. A, 40, 57-64.

Singh,Y.,Kamarujjama,M. and Khan,N.A. (2006). Integrals and Fourier series involving H-function, International J. of Mathematics and Analysis 1(1), 53-67.

Skibinski, P. (1970). Some expansion theorems for the H-function. Ann. Polon. Math. 23, 125-138.

Sneddon, I. N. (1957). Fourier Transform. McGraw-Hill, New York.

Sneddon, I.N. (1961). Special functions of mathematical physics and chemistry. 2nd ed. Oliver and Boyed, London.

Sneddon, I.N. (1972). The uses of Integral Transforms. McGraw-Hill, New York.

Srivastava, H.M. (1966). Some expansions in products of hypergeometric functions. Proc. Cambridge philos. Soc. 62, 245-247.

Srivastava, H.M. (1967). Generalized Newmann expansions involving hypergeometric functions. Proc. Cambridge philos. Soc. 63, 425-429.

Srivastava, H.M. (1972). A contour integral involving Fox' H-function. Indian J.Math. 14, 1-6.

Srivastava,H.M. (1972). Some formulas of Hermite and Carlitiz, Rev. Roumaine Math.Pures Appl.,17, 1257-1263.

Srivastava, H.M. and Buschman, R.G. (1972). Convolution integral equations with special function kernels. Wiley Eastern, New Delhi.

Srivastava, H.M. and Buschman, R.G. (1973). Composition of Fractional integral operators involving Fox's H-function. Acta Mexicana Ci. Tecn. 7, No. 1-2-3, 21-28.

Srivastava, H.M. and Karlsson, P.W.(1985), Multiple Gaussian Hypergeometric Series, Halssted Press (Ellis Horwood, Chichester),John Wiley and Sons, New York, Chechester and Toronto.

Srivastava, H.M., and Joshi, C.M. (1969). Integration of certain products associated with a generalized Meijer function. Proc. Cambridge philos. Soc. 65, 471-477.

Srivastava, H.M. and Panda, R. (1975). Some analytic or asymptotic confluent expansions for functions of several variables. Math. Comp. 29, 1115-1128.

Srivastava, H.M. and Panda, R. (1975a). Some expansion theorems and generating relations for the H-function of several complex variables. I. Comment. Math. Univ. St. paul 24, 119-137.

Srivastava, H.M. and Panda, R. (1976). Some expansion theorems and generating relations for the H-function of several complex variable. II. Comment. Math. Univ. St. paul 25, 169-197.

Srivastava, H.M. and Panda, R. (1976a). Some bilateral generating functions for a Class of generalized hypergeometric polynomials, J. Reine Angew. Math. 283/284, 265-274.

Srivastava, H.M. and Panda, R. (1976b). Expansion theorems for the H-function of several complex variable. J. Reine Angew. Math. 288, 129-145.

Srivastava, H.M. and Panda, R. (1978). Certain multidimensional integral transformations. I and II. Nederl. Akad. Wetensch. proc. Ser. A81=Indag. Math. 40, 118-131 and 132-144.

Srivastava, H.M. and Panda, R. (1979). Some multiple integral transformations involving the H-function of several variables. Nederl. Akad. Wetensch. Proc, A82=Indag. Math. 41, 353-362.

Srivastava, H.M. and Raina, R.K. (1981). New generating functions for certain polynomial systems associated with the H-function. Hokkaido Math. J. 10, 34-45.

Srivastava, H.M. and Singh, N.P. (1981). The integration of certain products or the multivariable H-function with a general class of polynomials (Abstract No. 81T-33-267). Amer. Math. Soc. Abstracts 2, 404-405.

Srivastava, H.M. and Singh, N.P. (1983). The integration of certain products or the multivariable H function with a general class of polynomials, Rend.circ.Math. Palermo, (Ser.2) 32,157-187.

Srivastava, H.M. and Srivastava, A. (1978). A New Class of orthogonal expansions for the H -function of several complex variables. Comment. Math. Univ. St. Paul 27, 59-69.

Srivastava, H.M., Gupta, K.C. and Goyal, S.P. (1982). The H -function of one and two variables with applications. South Asian Publishers, New Delhi and Madras.

Srivastava, H.M., Goyal, S.P. and Jain, R.N. (1990). Fractional integral operators involving a general class of polynomials. J. Math. Anal. Appl. 148(1). 87-100.

Srivastava, H.M., Koul, C.L. and Raina, R.K. (1981). A class of convolution integral. Univ. of Victoria (Reprint) No. DM- 243-IR.

Srivastava, R. (1981). Definite integrals associated with the H -function of several variables. Comment. Math. Unic. St. Paul 30, 1-5.

Suthar, D.L., Saxena, R. K. and Ram, J.(2004).Integral Formulas for the H-functions generalized fractional calculus, South East Asian J. Math . & Math. Sc. Vol. 3, pp. 69-74.

Swaroop, R. (1964). A study of Verma transform, collectanes Mathematica de Barcelona, Vol. 26, No. 1, 15-32.

Tandon, O.P. (1980). Contiguous relations for the H -function of n - variables. Indian J. pure Appl. Math. 11, 321-325.

Tandon, O.P. (1980a). Finite summation formulas for the H -function of n - variables. Indian J. pure Appl. Math. 11, 23-30.

Tandon, O.P. (1980b). Some finite series concerning the H -function of n - variables. Jnanabha (Professor Arthur Erdelyi Memorial Volume) 9 / 10.

Tsallis , C. (2003). in: Nonextensive Entropy: Interdisciplinary Applications, Eds. M. Gell-Mann, C. Tsallis, Oxford University Press, New York,1

Vasil'ev,S.S., Kocharov, G.E., Levkovskii , A.A. (1975). Izvestiya Akademii Nauk SSSR, Seriya Fizicheskaya 39, 310.

Vaishya, G.D., Jain, Renu and Verma, R.C. (1989). Certain properties of the I-function. Proc. Nat. Acad. Sci. India 59(A) II, 329-337.

Verma, A. (1966). A note on an expansion of hypergeometric functions of two variables. Math. Comp. 29, 413-417.

Verma, R.U. (1971). On the H-function of two variables. II. An. Sti. Univ. 'Al. I. Cuza' Iasi Sect. Ia Mat. (N.S.) 17, 103-109.

Weyl, H. (1917). Bemerkung zum Begriff des Differential quotienten gebrochener ordenung, Viertel jahrsschrifted. Naturf. Gesellschaft, in Zurich, 62, 296-302.

Whittaker, E.T. and Watson, G.N. (1964). A Course of Modern Analysis, 4th ed. Cambridge.

Yadav, R.K. and Purohit, S.D. (2008). On generalized weyl fractional q-integral operator involving generalized basic hypergeometric function, fractional Calculus and Applied Analysis, vol. 11(2),1-14.

Yadav, R.K. and Purohit, S.D. (2006). On applications of Weyl fractional q-integral operator to generalized basic hypergeometric functions, Kyungpook Math. J. 46: 235-245.

◎编辑手记

世界著名数学家 L. Felix 曾指出：

> 我们生活在伟大的数学思想更新时期，这是不可估量的好运气. 如果不从提供给我们的大量新鲜空气中获益，不是再次发现青春与热情，那就与我们的使命相违背了. 因为在我们被邀请来回顾价值时，数学正在打扮得年轻、貌美，变得更加有用、更加丰产.

本书是一部英文版的数学专著，中文书名或可译为《与广义积分变换有关的分数次演算：对分数次演算的研究》.

本书的作者为哈门德拉·库马尔·曼迪亚(Harmendra Kumar Mandia)，印度人，现为印度拉贾斯坦大学数学系教授;还有一位为亚什万特·辛格(Yashwant Singh),印度人,塞思·莫特勒学院数学系博士生导师.

本书的主要内容如下:

在第1章中,作者简要考查了分数次演算、分数次积分算子和广义超几何函数,给出了带一个或多个变量的 I—函数、G—函数、H—函数和 A—函数的定义.其中一些将用于呈现后续结果.

第2章第1部分中作者建立了二重 Laplace 变换和二重 Hankel 变换的关系.

下面的公式是证明中需要的

$$\int_0^\infty\int_0^\infty x^{s-1}y^{t-1}H[ax^\lambda,by^\mu]\mathrm{d}x\mathrm{d}y$$

$$=\frac{a^{-\frac{s}{\lambda}}b^{-\frac{t}{\mu}}}{\lambda\mu}\phi\left(-\frac{s}{\lambda},-\frac{t}{\mu}\right)\theta_2\left(-\frac{s}{\lambda}\right)\theta_3\left(-\frac{t}{\mu}\right)$$

带两个变量的 H—函数对 Srivastava 和 Panda 的基本定义和表示是

$$H[x,y]$$

$$=H\begin{bmatrix}x\\y\end{bmatrix}$$

$$=H^{0,n_1:m_2,n_2:m_3,n_3}_{p_1,q_1:p_2,q_2:p_3,q_3}\left[\begin{matrix}x\\y\end{matrix}\left|\begin{matrix}(a_j;\alpha_j,A_j)_{1,p_1}:(c_j,\gamma_j)_{1,p_2},(e_j,E_j)_{1,p_3}\\(b_j;\beta_j,B_j)_{1,q_1}:(d_j,\delta_j)_{1,q_2},(f_j,F_j)_{1,q_3}\end{matrix}\right.\right]$$

$$=-\frac{1}{4\pi^2}\int_{L_1}\int_{L_2}\phi(\xi,\eta)\theta_2(\xi)\theta_3(\eta)x^\xi y^\eta\mathrm{d}\xi\mathrm{d}\eta$$

第2部分中作者得到了二重 Laplace 变换和二重 Mellin 变换之间的关系.有一个或两个变量的 H—函数的乘积的二重 Laplace-Mellin 变换后面可

以得到.

如果 $F(p_1,p_2)$ 是 Laplace 变换,且 $M(p_1,p_2)$ 是 $f(t_1,t_2)$ 的 Mellin 变换,那么

$$F(p_1,p_2)=\sum_{s_1=0}^{\infty}\sum_{s_2=0}^{\infty}\frac{(-p_1)^{s_1}}{s_1!}\frac{(-p_2)^{s_2}}{s_2!}M(s_1+1,s_2+1)$$

第 3 章第 1 部分中作者根据有两个变量的 Whittaker 变换得到了一个新定理. 该结果通过应用 Weyl 型的二维Erdelyi- Kober 算子得到. 一些已知的和新的特殊例子在最后给出.

设

$$g(p,q)=W_{\lambda_1,\mu_1}^{\lambda,\mu}[F(x,y);\rho,\sigma,p,q]=\int_b^{\infty}\int_d^{\infty}(px)^{\rho-1}(qy)^{\sigma-1}\cdot$$

$$\exp\left(\frac{1}{2}px+\frac{1}{2}qy\right)W_{\lambda,\mu}(px)W_{\lambda_1,\mu_1}(qy)F(x,y)\mathrm{d}x\mathrm{d}y$$

是二维 Whittaker 变换,满足 $\alpha>0,\beta>0$,下面的结果成立

$$K_p^{\eta,\alpha}K_q^{\delta,\beta}[g(p,q)]=G_{\lambda_1,\beta,\delta,\mu_1}^{\lambda,\alpha,\eta,\mu}[F(x,y);\rho,\sigma,p,q]$$

二维 Whittaker 变换可以化为一个二维 Laplace 变换,且我们有一个 Saxena 等人给出的结果.

第 2 部分中作者导出了基本类比 $\overline{H}$— 函数的一个展开公式, 通过两个函数乘积的 q 导子型的 q-Leibniz 规则的应用得到. 展开公式涉及 Fox 的 H— 函数的一个基本类比,Meijer 的 G— 函数和 MacRobert 的 E— 函数是由主要结果的特殊例子导出的.

作者得到的主要结果如下

$$\overline{H}_{p+1,q+1}^{m+1,n}\left[\rho(zq^{\mu})^{k};q\,\middle|\,\begin{matrix}A^{*},(\lambda,k)\\(\mu+\lambda,k),B^{*}\end{matrix}\right]$$

$$=\sum_{R=0}^{\mu}\frac{(-1)^{R}q^{\frac{R(R+1)}{2}+\lambda R}(q^{-\mu};q)_{R}(q^{\lambda};q)_{\mu-R}}{(q;q)_{R}}\cdot$$

$$\overline{H}_{p+1,q+1}^{m+1,n}\left[\rho(zq^{\mu})^{k};q\,\middle|\,{}^{A^{*},(0,k)}_{(R,k),B^{*}}\right]$$

在第4章第1部分中,作者得到了广义微分算子 $D_{k,\alpha,x}^{m}$ 的两个新的有用的定理. 作为主要结果的一个应用,作者获得了 $\overline{H}$ — 函数的两个乘法公式.

定理1

$$D_{l,\lambda-\mu,t}^{m}\{t^{\lambda-1}S_{N}^{M}[wt^{\rho}]f(xt)\}$$
$$=\sum_{k=0}^{\left[\frac{N}{M}\right]}\frac{(-N)_{Mk}}{k!}A_{n,k}w^{k}\sum_{n=0}^{\infty}\frac{(-x)^{n}}{n!}\cdot$$
$$\prod_{p=0}^{m-1}\frac{\Gamma(\lambda+\rho k+pl)}{\Gamma(\mu+\rho k+pl)}t_{m+1}^{\lambda+\rho k+ml-1}\cdot$$
$$F_{m}\left[{}^{-n,\lambda+\rho k,\lambda+\rho k+l,\cdots,\lambda+\rho k+(m-1)l}_{\mu+\rho k,\mu+\rho k+l,\cdots,\mu+\rho k+(m-1)l}\right]D_{x}^{n}\{f(x)\}$$

定理2

$$D_{l,\lambda-\mu,t}^{m}\{t^{\lambda}S_{N}^{M}[wt^{\rho}]f(xt)\}=\sum_{k=0}^{\left[\frac{N}{M}\right]}\frac{(-N)_{Mk}}{k!}A_{n,k}w^{k}\cdot$$
$$\sum_{n=0}^{\infty}\frac{(-x)^{n}}{n!}\prod_{p=0}^{m-1}\frac{\Gamma(\lambda+\rho k+pl)\Gamma(1-\mu-\rho k-pl)_{n}}{\Gamma(\mu+\rho k+pl)\Gamma(1-\lambda-\rho k-pl)_{n}}$$
$$t_{m+1}^{\lambda+\rho k+ml-1}F_{m}\left[{}^{-n,\lambda+\rho k-n,\lambda+\rho k-n+l,\cdots,\lambda+\rho k-n+(m-1)l}_{\mu+\rho k-n,\mu+\rho k-n+l,\cdots,\mu+\rho k-n+(m-1)l}\right]D_{x}^{n}\{f(x)\}$$

在第4章第2部分中,作者得到了 I— 函数的二重有限级数的四个变换公式. 这些公式后来又被用于得到 $\overline{H}$ — 函数的二重求和公式. 作者的结果的性质是相当普遍的,并且很多求和公式可以被演绎成特殊的例子.

第一个公式

$$\sum_{m,n=0}^{\infty}x^{m}y^{n}I_{p_i+2,q_i+1:r}^{m,n+2}\left[z\,\middle|\,{}^{(1-a-m,\rho),(1-b-n,\sigma),A^{*}}_{B^{*},(1-a-b-m-n,\sigma+\rho)}\right]$$
$$=(x+y-xy)^{-1}\cdot$$
$$\left\{x^{s+1}\sum_{s=0}^{\infty}x^{s+1}I_{p_i+2,q_i+1:r}^{m,n+2}\left[z\,\middle|\,{}^{(1-a-s,\rho),(1-b,\sigma),A^{*}}_{B^{*},(1-a-b-s,\sigma+\rho)}\right]+\right.$$

$$\sum_{t=0}^{\infty} y^{t+1} I_{p_i+2,q_i+1:r}^{m,n+2}\left[z \,\middle|\, {}^{(1-a,\rho),(1-b-t,\sigma),A^*}_{B^*,(1-a-b-t,\sigma+\rho)}\right]\Big\}$$

第二个公式

$$\sum_{m,n=0}^{\infty} \frac{x^m y^n}{m!\ n!} I_{p_i,q_i;r}^{m,n}\left[z \,\middle|\, {}^{(1-a-m-n,u),(1-b-m,v),A^*}_{B^*,(1-c-m,\omega)}\right]$$

$$=\sum_{k=0}^{\infty} \frac{1}{k!}(1-y)^{-a}\left(\frac{x}{1-y}\right)^{k}\cdot$$

$$I_{p_i+2,q_i+1:r}^{m,n+2}\left[z(1-y)^{-u} \,\middle|\, {}^{(1-a-k,u),(1-b-k,v),A^*}_{B^*,(1-c-k,\omega)}\right]$$

第三个公式

$$\sum_{m,n=0}^{\infty} \frac{x^m y^n}{m!\ n!} I_{p_i+3,q_i+2:r}^{m,n+3}\left[z \,\middle|\, {}^{(1-a-m-n,u),(1-b-m,v),(1-b'-n,\omega)A^*}_{B^*,(1-a-m,u),(1-a-n,u)}\right]$$

$$=\sum_{k=0}^{\infty} \frac{1}{k!}(1-x)^{-b}(1-y)^{b'}\left(\frac{xy}{(1-x)(1-y)}\right)^{k}\cdot$$

$$I_{p_i+2,q_i+1:r}^{m,n+2}\left[z(1-x)^{-v}(1-y)^{-u} \,\middle|\, {}^{(1-b-k,v),(1-b'-k,\omega),A^*}_{B^*,(1-a-k,u)}\right]$$

第四个公式

$$\sum_{m,n=0}^{\infty} \frac{x^m y^n}{m!\ n!} I_{p_i+3,q_i+1:r}^{m,n+3}\left[z \,\middle|\, {}^{(1-a-m-n,u),(1-b-m,v),(1-b'-n,\omega)A^*}_{B^*,(1-b-b'-m-n,\omega+v)}\right]$$

$$=\sum_{k=0}^{\infty}(1-y)^{-a}\frac{1}{K!}\left(\frac{x-y}{1-y}\right)^{k}\cdot$$

$$I_{p_i+3,q_i+1:r}^{m,n+3}\left[z(1-y)^{-u} \,\middle|\, {}^{(1-a-k,u),(1-b-k,v),(1-b',\omega),A^*}_{B^*,(1-b-b'-k,\omega+v)}\right]$$

在第5章第1部分中,作者得到了表示相关函数的图像和原象之间的相互联系的四个有趣的定理,也给出了这些定理的六个推论,之后,通过应用这些定理,得到了五个新的通积分.

这部分中给出的I—函数被定义并表示成

$$I[z]=I_{p_i,q_i:r}^{m,n}[z]=I_{p_i,q_i:r}^{m,n}\left[z \,\middle|\, {}^{(a_j,\alpha_j)_{1,n},(a_{ji},\alpha_{ji})_{n+1,p_i}}_{(b_j,\beta_j)_{1,m},(b_{ji},\beta_{ji})_{m+1,q_i}}\right]$$

$$=\frac{1}{2\pi\omega}\int_L \phi(\xi)z^{\xi}\mathrm{d}\xi$$

在第5章第2部分中,作者给出了一个卷积积分函数的解,该函数的核心是一个广义超几何函数和两个变量的 H－函数,并讨论了主要结果的有趣的特殊例子.

下面这个卷积积分函数的解已经给出

$$\int_0^x (x-t)^{\sigma-1}\ {}_PF_Q[(g_P);(h_Q);a(x-t)^\eta]\cdot$$

$$H_{p_1,q_1:p_2,q_2:p_3,q_3}^{o,n_1:m_2,n_2:m_3,n_3}\left[\begin{matrix}(x-t)\\(x-t)\end{matrix}\middle|\begin{matrix}(a_j;\alpha_j,A_j)_{1,p_1}:(c_j,\gamma_j)_{1,p_2}:(e_j,E_j)_{1,p_3}\\(b_j;\beta_j,B_j)_{1,q_1}:(d_j,\delta_j)_{1,q_2}:(f_j,F_j)_{1,q_3}\end{matrix}\right]\cdot$$

$$f(t)\mathrm{d}t=g(x)$$

在第6章第1部分中,作者推导了与 I－函数乘积有关的三个分数阶积分算子的组合和由 Srivastava 得到的一类多项式

下面的积分方程可以被导出

$$\mathrm{Y}_\alpha^h\{f(x)\}=x^{-h-\alpha}\mathrm{A}_\alpha\{x^h f(x)\}$$

$$=\frac{x^{-h-\alpha}}{\Gamma(\alpha)}\int_0^x (x-s)^{\alpha-1} I_{p_i,q_i:r}^{m,n}\left[z\left(1-\frac{s}{x}\right)^c\right]\cdot$$

$$\mathrm{S}_b^a\left[y\left(1-\frac{s}{x}\right)^\sigma\right]s^h f(s)\mathrm{d}s$$

在第6章第2部分中,作者计算了一个涉及指数函数、正弦函数、广义超几何级数和 I－函数的积分,并且利用它计算了一个二重积分,建立了广义超几何函数乘积的 Fourier 级数. 作者也导出了一个 I－函数的二重 Fourier 指数级数. 作者的结果在本质上是统一的,并可以作为关键公式,从中可以得出许多特殊情况的结果.

下面的公式在证明中可得

$$\int_0^\pi \sin x)^{\omega-1}\mathrm{e}^{imx}\ {}_PF_Q\left[\begin{matrix}\alpha_P\\\beta_Q\end{matrix};c(\sin x)^{2h}\right]\mathrm{d}x$$

$$=\frac{\pi e^{\frac{im\pi}{2}}}{2^{\omega-1}}\sum_{r=0}^{\infty}\frac{(\alpha_P)_r c^r\Gamma(\omega+2hr)}{(\beta_Q)_r r!\ 2^{2hr}\Gamma\left(\frac{\omega+2hr\pm m+1}{2}\right)}$$

在第 7 章中，作者展示了一个对天体物理热核函数积分的解析证明，该函数是在 Boltzmann-Gibbs 统计力学的基础上导出的.

Mellin-Barnes 指数函数的积分表示，即

$$\exp[-x]=\frac{1}{2\pi\omega}\int_L\Gamma(s)x^{-s}\mathrm{d}s,\ |x|<\infty$$

本书的目录如下：

第6章　(1) 关于广义超几何函数的乘积的广义分数次积分的组成部分

(2) 关于广义超几何函数的一些积分和 Fourier 级数

第7章　对 Boltzmann-Gibbs 统计学、Tsallis 统计学和 $\overline{H}$ — 函数的天体物理热核函数的统一研究

在中文系有一门很有意思的课，叫“中外比较文学”，据说是北京大学的乐黛云先生倡导的，我们不妨借用一下，将近年来我国学者对本书相关主题的研究的情况向读者们汇报一下. 1992 年沈克精教授以《梅林(Mellin) 变换及其应用》① 为题从 Mellin 变换的定义出发导出其反演定理及其性质，并用此变换求解对偶积分方程，如下：

1　定义及性质

定义　设 $p=\sigma+\mathrm{i}s(\sigma,s\in\mathbf{R})$，$f(x)x^{\sigma-1}\in L_1(0,+\infty)$，
那么，当 $f(x)x^{p-1}\in L_1(0,+\infty)$ 时，有

$$F(p)=\int_0^{+\infty}f(x)x^{p-1}\mathrm{d}x \tag{1}$$

称为 $f(x)$ 的 Mellin 变换.

由上述定义，可得到 Mellin 反演定理.

定理　若对于某个 $k>0$，$\int_0^{+\infty}f(x)x^{k-1}\mathrm{d}x$ 有

① 摘自《安徽大学学报》(自然科学版)，1992(3)：15-22.

界,并设

$$F(p)=\int_{0}^{+\infty} f(x)x^{p-1}\mathrm{d}x$$

则

$$f(x)=\frac{1}{2\pi \mathrm{i}}\int_{\sigma-\mathrm{i}\infty}^{\sigma+\mathrm{i}\infty} F(p)x^{-p}\mathrm{d}p \qquad (2)$$

证　令 $x=\mathrm{e}^{y}, y\in(-\infty,+\infty)$,考虑 $f(\mathrm{e}^{y})\mathrm{e}^{\sigma y}, y\in(-\infty,+\infty)$,必有

$$\int_{-\infty}^{+\infty} |f(\mathrm{e}^{y})| \mathrm{e}^{\sigma y}\mathrm{d}y=\int_{0}^{+\infty} |f(x)| x^{\sigma-1}\mathrm{d}x<+\infty$$

故此,这个函数的 Fourier 变换为

$$\begin{aligned}\overline{F}(s)&=\frac{1}{\sqrt{2\pi}}\int_{-\infty}^{+\infty} f(\mathrm{e}^{y})\mathrm{e}^{\sigma y}\mathrm{e}^{\mathrm{i}sy}\mathrm{d}y\\&=\frac{1}{\sqrt{2\pi}}\int_{0}^{+\infty} f(x)x^{p-1}\mathrm{d}x\\&=\frac{1}{\sqrt{2\pi}}F(p)\end{aligned}$$

为此,$f(x)$ 的 Mellin 变换 $F(p)$ 恰是$\sqrt{2\pi}\,\overline{F}(s)$,即为$\sqrt{2\pi}\,f(\mathrm{e}^{y})\mathrm{e}^{\sigma y}$ 的 Fourier 变换,故得

$$\sqrt{2\pi}\,f(\mathrm{e}^{y})\mathrm{e}^{\sigma y}=\frac{1}{\sqrt{2\pi}}\int_{-\infty}^{+\infty}\overline{F}(s)\mathrm{e}^{-\mathrm{i}sy}\mathrm{d}s$$

$$f(\mathrm{e}^{y})\mathrm{e}^{\sigma y}=\frac{1}{2\pi}\int_{-\infty}^{+\infty}\overline{F}(s)\mathrm{e}^{-\mathrm{i}sy}\mathrm{d}s$$

$$f(x)=\frac{1}{2\pi \mathrm{i}}\int_{\sigma-\mathrm{i}\infty}^{\sigma+\mathrm{i}\infty}\overline{F}\left(\frac{p-\sigma}{\mathrm{i}}\right)\mathrm{e}^{-py}\mathrm{d}p$$

记 $F(p)=\overline{F}\left(\frac{p-\sigma}{\mathrm{i}}\right)$,即得

$$f(x)=\frac{1}{2\pi \mathrm{i}}\int_{\sigma-\mathrm{i}\infty}^{\sigma+\mathrm{i}\infty} F(p)x^{-p}\mathrm{d}p$$

由上述定义,可得下列两个基本性质:

（Ⅰ）各阶导数的 Mellin 变换.

$$\int_0^{+\infty}\frac{\mathrm{d}^r f}{\mathrm{d}x}x^{p-1}\mathrm{d}x=\left[\frac{\mathrm{d}^{r-1}}{\mathrm{d}x^{r-1}}\right]_0^{+\infty}-(p-1)\int_0^{+\infty}\frac{\mathrm{d}^{r-1}f}{\mathrm{d}x^{r-1}}x^{p-2}\mathrm{d}x \tag{3}$$

如果假设使得上式括号化为零的函数，那么，便得到下面的关系式

$$F^{(r)}(p)=-(p-1)F(r-1)(p-1) \tag{4}$$

其中

$$F^{(r)}(p)=\int_0^{+\infty}\frac{\mathrm{d}^r f}{\mathrm{d}x^r}x^{p-1}\mathrm{d}x$$

重复应用上述规则，便可得到

$$F^{(r)}(p)=(-1)^r\frac{\Gamma(p)}{\Gamma(p-r)}F(p-r)$$

（Ⅱ）乘积的 Mellin 变换.

设 $F_1(p),F_2(p)$ 分别为 $f_1(x),f_2(x)$ 的 Mellin 变换，则 $f_1(x)f_2(x)$ 的 Mellin 变换为

$$\begin{aligned}&\int_0^{+\infty}f_1(x)f_2(x)x^{p-1}\mathrm{d}x\\&=\int_0^{+\infty}f_2(x)x^{p-1}\mathrm{d}x\,\frac{1}{2\pi\mathrm{i}}\int_{\sigma-\mathrm{i}\infty}^{\sigma+\mathrm{i}\infty}F_1(t)x^{-t}\mathrm{d}t\\&=\frac{1}{2\pi\mathrm{i}}\int_{\sigma-\mathrm{i}\infty}^{\sigma+\mathrm{i}\infty}F_1(t)\mathrm{d}t\int_0^{+\infty}f_2(x)x^{p-t-1}\mathrm{d}x\end{aligned}$$

即

$$\int_0^{+\infty}f_1(x)f_2(x)x^{p-1}\mathrm{d}x=\frac{1}{2\pi\mathrm{i}}\int_{\sigma-\mathrm{i}\infty}^{\sigma+\mathrm{i}\infty}F_1(t)F_2(p-t)\mathrm{d}t \tag{5}$$

上式的特殊情形为

$$\begin{aligned}&\int_0^{+\infty}f_1(x)f_2(x)\mathrm{d}x\\&=\frac{1}{2\pi\mathrm{i}}\int_0^{+\infty}f_2(x)\mathrm{d}x\int_{\sigma-\mathrm{i}\infty}^{\sigma+\mathrm{i}\infty}F_1(p)x^{-p}\mathrm{d}p\end{aligned}$$

$$=\frac{1}{2\pi\mathrm{i}}\int_{\sigma-\mathrm{i}\infty}^{\sigma+\mathrm{i}\infty}F_1(p)\mathrm{d}p\int_0^{+\infty}f_2(x)x^{(p-1)-1}\mathrm{d}x$$

$$=\frac{1}{2\pi\mathrm{i}}\int_{\sigma-\mathrm{i}\infty}^{\sigma+\mathrm{i}\infty}F_1(p)F_2(1-p)\mathrm{d}p \tag{6}$$

由此，又可得乘积 $F_1(p)F_2(p)$ 的 Mellin 反变换式为

$$\frac{1}{2\pi\mathrm{i}}\int_{\sigma-\mathrm{i}\infty}^{\sigma+\mathrm{i}\infty}F_1(p)F_2(p)x^{-p}\mathrm{d}p -$$

$$\frac{1}{2\pi\mathrm{i}}\int_{\sigma-\mathrm{i}\infty}^{\sigma+\mathrm{i}\infty}F_1(p)x^{-p}\mathrm{d}p\int_0^{+\infty}f_2(u)u^{p-1}\mathrm{d}u$$

$$=\int_0^{+\infty}f_2(u)\frac{\mathrm{d}u}{u}\frac{1}{2\pi\mathrm{i}}\int_{\sigma-\mathrm{i}\infty}^{\sigma+\mathrm{i}\infty}F_1(p)\left(\frac{x}{u}\right)^{-p}\mathrm{d}p$$

$$=\int_0^{+\infty}f_1\left(\frac{x}{u}\right)f_2(u)\frac{\mathrm{d}u}{u} \tag{7}$$

特别地

$$\frac{1}{2\pi\mathrm{i}}\int_{\sigma-\mathrm{i}\infty}^{\sigma+\mathrm{i}\infty}F_1(p)F_2(p)\mathrm{d}p$$

$$=\frac{1}{2\pi\mathrm{i}}\int_{\sigma-\mathrm{i}\infty}^{\sigma+\mathrm{i}\infty}F_1(p)\mathrm{d}p\int_0^{+\infty}f_2(u)u^{p-1}\mathrm{d}u$$

$$=\int_0^{+\infty}f_1\left(\frac{1}{u}\right)f_2(u)\frac{\mathrm{d}u}{u} \tag{8}$$

2　利用 Mellin 变换求解对偶积分方程

给定

$$\begin{cases}\displaystyle\int_0^{+\infty}y^{\alpha}f(y)J_r(xy)\mathrm{d}y=g(x),0<x<1\\ \displaystyle\int_0^{+\infty}f(y)J_r(xy)\mathrm{d}y=0,x>1\end{cases} \tag{9}$$

其中 $f(y)$ 为待定函数，$g(x)$ 为已知函数.

形如(9)的方程,称为对偶积分方程,可以用如下方法求解.

函数 $y^{\alpha}J_r(xy)$ 的 Mellin 变换式是

$$J_\alpha(p)=\int_0^{+\infty}y^{\alpha+p-1}J_r(xy)\mathrm{d}y=\frac{2^{\alpha+p-1}}{x^{\alpha+p}}\frac{\Gamma\left(\frac{\alpha+r+p}{2}\right)}{\Gamma\left(\frac{2-\alpha-p+r}{2}\right)}$$

上式中令 $\alpha=0$,便可推出 Bessel 函数 $J_r(xy)$ 的 Mellin 变换式是

$$J_r(p)=\frac{2^{p-1}}{x^{p}}\frac{\Gamma\left(\frac{r+p}{2}\right)}{\Gamma\left(\frac{2-p+r}{2}\right)}$$

由 Mellin 变换式的基本性质(Ⅱ)的特殊情形,方程(9)可得

$$\int_0^{+\infty}y^{\alpha}f(y)J_r(xy)\mathrm{d}y=\frac{1}{2\pi\mathrm{i}}\int_{\sigma-\mathrm{i}\infty}^{\sigma+\mathrm{i}\infty}F(p)J_\alpha(1-p)\mathrm{d}p$$

$$\int_0^{+\infty}f(y)J_r(xy)\mathrm{d}y=\frac{1}{2\pi\mathrm{i}}\int_{\sigma-\mathrm{i}\infty}^{\sigma+\mathrm{i}\infty}F(p)J_\alpha(1-p)\mathrm{d}p$$

其中 $F(p)$ 是 $f(y)$ 的 Mellin 变换式.

因此,对偶积分方程(9)便与下面一对方程等价

$$\begin{cases}\dfrac{1}{2\pi\mathrm{i}}\displaystyle\int_{\sigma-\mathrm{i}\infty}^{\sigma+\mathrm{i}\infty}F(p)\dfrac{2^{\alpha-p}\Gamma\left(\frac{\alpha+r+1-p}{2}\right)}{x^{1+\alpha-p}\Gamma\left(\frac{1-\alpha+p+r}{2}\right)}\mathrm{d}p=g(x),\\ \qquad 0<x<1\\ \dfrac{1}{2\pi\mathrm{i}}\displaystyle\int_{\sigma-\mathrm{i}\infty}^{\sigma+\mathrm{i}\infty}F(p)\dfrac{2^{-p}\Gamma\left(\frac{r+1-p}{2}\right)}{x^{1-p}\Gamma\left(\frac{r+1+p}{2}\right)}\mathrm{d}p=0,x>1\end{cases}\tag{10}$$

若令

$$F(p)=2^{p-d}\frac{\Gamma\left(\frac{1+r+p}{2}\right)}{\Gamma\left(\frac{1+r+\alpha-p}{2}\right)}\chi(p)$$

则上列方程便化为

$$\begin{cases}\frac{1}{2\pi i}\int_{\sigma-i\infty}^{\sigma+i\infty}\frac{\Gamma\left(\frac{1+\alpha+p}{2}\right)}{\Gamma\left(\frac{1+r-\alpha+p}{2}\right)}\chi(p)x^{p-1-\alpha}dp=g(x), \\ \qquad 0<x<1 \\ \frac{1}{2\pi i}\int_{\sigma-i\infty}^{\sigma+i\infty}\frac{\Gamma\left(\frac{1+r-p}{2}\right)}{\Gamma\left(\frac{1+r+\alpha-p}{2}\right)}\chi(p)x^{p-1}dp=0, x>1\end{cases} \tag{11}$$

将式(11)中的第一式乘以 $x^{\alpha-\omega}$，其中 $R(p)-R(\omega)>0$（R 代表实数部分），并对 x 由 0 到 1 求积分，则得下列关系式

$$\frac{1}{2\pi i}\int_{\sigma-i\infty}^{\sigma+i\infty}\frac{\Gamma\left(\frac{1+r+p}{2}\right)}{\Gamma\left(\frac{1+r-\alpha+p}{2}\right)}\frac{\chi(p)}{p-\omega}dp=G(\alpha-\omega+1)$$

其中 $G(p)=\int_0^1 g(x)x^{p-1}dp$ 是下面函数的 Mellin 变换式

$$y(x)=\begin{cases}g(x), 0\leqslant x\leqslant 1\\ 0, x>1\end{cases}$$

将积分路线由 $R(p)=\sigma$ 移到 $R(p)=\sigma'<R(\omega)$，如图 1 所示，则得

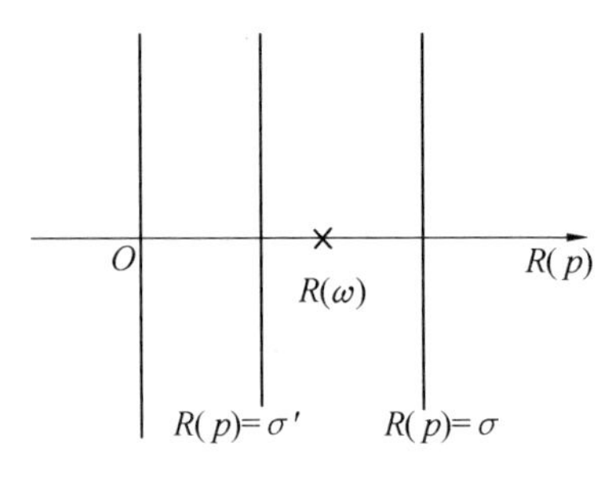

图 1

$$\frac{1}{2\pi i}\int_{\sigma'-i\infty}^{\sigma'+i\infty}\frac{\Gamma\left(\frac{1+r+p}{2}\right)}{\Gamma\left(\frac{1+r-\alpha+p}{2}\right)}\frac{\chi(p)}{p-\omega}dp+$$

$$\frac{\Gamma\left(\frac{1+r+\omega}{2}\right)\chi(\omega)}{\Gamma\left(\frac{1+r-\alpha+\omega}{2}\right)}=G(\alpha-\omega+1)$$

但此式左边的积分当 $R(\omega)>\sigma'$ 时，是 ω 的解析函数，所以函数

$$G(\alpha-\omega+1)-\frac{\Gamma\left(\frac{1+r+\omega}{2}\right)\chi(\omega)}{\Gamma\left(\frac{1+r-\alpha+\omega}{2}\right)}$$

当 $R(\omega)>\sigma'$ 时，也是 ω 的解析函数，因此

$$\chi(\omega)-\frac{\Gamma\left(\frac{1+r-\alpha+\omega}{2}\right)}{\Gamma\left(\frac{1+r+\omega}{2}\right)}G(\alpha-\omega+1)$$

也是一样，从而

$$\frac{1}{2\pi i}\int_{\sigma-i\infty}^{\sigma+i\infty}\left[\chi(p)-\frac{\Gamma\left(\frac{1+r-\alpha+p}{2}\right)}{\Gamma\left(\frac{1+r+p}{2}\right)}G(\alpha+1-p)\right]\frac{dp}{p-\omega}=0$$

其中 $R(\omega)<\sigma$.

仿照以上方法,由(11)的第二式,可以推出

$$\frac{1}{2\pi i}\int_{\sigma-i\infty}^{\sigma+i\infty}\frac{\chi(p)}{p-\omega}dp=\chi(\omega),R(\omega)<\sigma$$

于是

$$\chi(\omega)=\frac{1}{2\pi i}\int_{\sigma-i\infty}^{\sigma+i\infty}\frac{\Gamma\left(\frac{1+r-\alpha+p}{2}\right)}{\Gamma\left(\frac{1+r+p}{2}\right)}\frac{G(\alpha-p+1)}{p-\omega}dp$$

由定义可知

$$G(\alpha-p+1)=\int_0^1 g(\zeta)\zeta^{\alpha-p}d\zeta$$

且有

$$\frac{1}{p-\omega}=\int_0^1\eta^{p-\omega-1}d\eta$$

将积分交换次序便得到

$$\chi(\omega)=\int_0^1 g(\zeta)\zeta^{\alpha}d\zeta\int_0^1\eta^{-\omega-1}d\eta\,\frac{1}{2\pi i}\int_{\sigma-i\infty}^{\sigma+i\infty}\frac{\Gamma\left(\frac{1+r-\alpha+p}{2}\right)}{\Gamma\left(\frac{1+r+p}{2}\right)}\left(\frac{\zeta}{\eta}\right)^{-p}dp$$

考虑积分

$$I=\int_0^1 x^{2\beta}(1-x^2)^{\lambda-1}x^{p-1}dx$$

利用积分变换,令 $\zeta=x^2$ 可得

$$I=\frac{1}{2}\int_0^1\zeta^{\beta+\frac{p}{2}-1}(1-\zeta)^{\lambda-1}\mathrm{d}\zeta=\frac{\Gamma(\lambda)\Gamma\left(\beta+\frac{p}{2}\right)}{2\Gamma\left(\lambda+\beta+\frac{p}{2}\right)}$$

由 Mellin 反演定理,可得

$$\frac{1}{2\pi\mathrm{i}}\int_{\sigma-\mathrm{i}\infty}^{\sigma+\mathrm{i}\infty}\frac{\Gamma\left(\beta+\frac{p}{2}\right)}{\Gamma\left(\lambda+\beta+\frac{p}{2}\right)}x^{-p}\mathrm{d}p$$

$$=\begin{cases}\dfrac{2x^{2\beta}(1-x^2)^{\lambda-1}}{\Gamma(\lambda)},0\leqslant x\leqslant 1\\ 0,x>1\end{cases}$$

在上述结果中,令 $x=\dfrac{\zeta}{\eta}$,$\beta=\dfrac{1}{2}(1+r-\alpha)$,$\lambda=\dfrac{\alpha}{2}$,则得积分式

$$\frac{1}{2\pi\mathrm{i}}\int_{\sigma-\mathrm{i}\infty}^{\sigma+\mathrm{i}\infty}\frac{\Gamma\left(\frac{1+r+p-\alpha}{2}\right)}{\Gamma\left(\frac{1+r+p}{2}\right)}\left(\frac{\zeta}{\eta}\right)^{-p}\mathrm{d}p$$

$$=\begin{cases}\dfrac{2}{\Gamma\left(\frac{\alpha}{2}\right)}\zeta^{1+r-\alpha}(\eta^2-\zeta^2)^{\frac{\alpha}{2}-1}\eta^{1-r},\eta\geqslant\zeta\\ 0,0<\eta<\zeta\end{cases}$$

从而

$$\chi(\omega)=\frac{2}{\Gamma\left(\frac{\alpha}{2}\right)}\int_0^1 g(\zeta)\zeta^{1+r}\mathrm{d}\zeta\int_\zeta^1\eta^{-\omega-r}(\eta^2-\zeta^2)^{\frac{\alpha}{2}-1}\mathrm{d}\eta$$

上述积分,调换积分顺序(图 2),则得

$$\chi(\omega)=\frac{2}{\Gamma\left(\frac{\alpha}{2}\right)}\int_0^1\eta^{-\omega-r}\mathrm{d}\eta\int_0^\eta g(\zeta)\zeta^{1+r}(\eta^2-\zeta^2)^{\frac{\alpha}{2}-1}\mathrm{d}\zeta$$

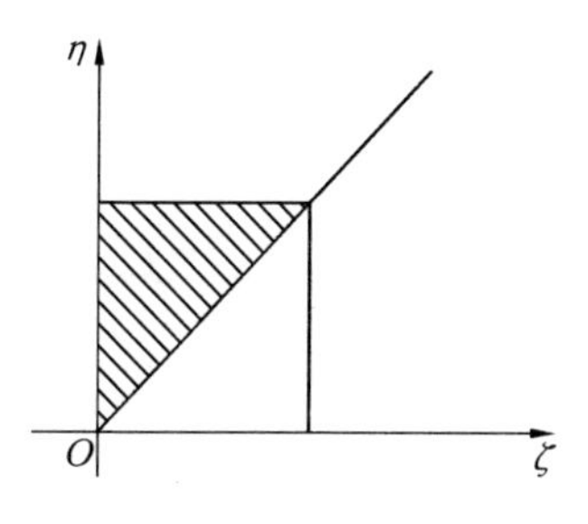

图 2

$$=\frac{2}{\Gamma\left(\frac{\alpha}{2}\right)}\int_0^1 \eta^{\alpha-\omega}\mathrm{d}\eta\int_0^1 g(\eta\delta)\delta^{r+1}(1-\delta^2)^{\frac{\alpha}{2}-1}\mathrm{d}\delta$$

但由 $\chi(p)$ 的定义及 Mellin 反演定理

$$f(x)=\frac{1}{2\pi \mathrm{i}}\int_{\sigma-\mathrm{i}\infty}^{\sigma+\mathrm{i}\infty}F(p)x^{-p}\mathrm{d}p$$

可求出函数 $f(x)$ 的表达式如下

$$f(x)=\frac{2}{\Gamma\left(\frac{\alpha}{2}\right)}\int_0^1 \eta^{\alpha}\mathrm{d}\eta\int_0^1 g(\eta\delta)\delta^{r+1}(1-\delta^2)^{\frac{\alpha}{2}-1}\mathrm{d}\delta_{\lambda}\cdot$$

$$\frac{1}{2\pi \mathrm{i}}\int_{\sigma-\mathrm{i}\infty}^{\sigma+\mathrm{i}\infty}2^{p-\alpha}(x\eta)^{-p}\ \frac{\Gamma\left(\frac{1+r+p}{2}\right)}{\Gamma\left(\frac{1+r-\alpha-p}{2}\right)}\mathrm{d}p$$

$J_{r+\frac{\alpha}{2}}(at)t^{1-\frac{\alpha}{2}}$ 的 Mellin 变换为

$$\int_0^{+\infty}J_{r+\frac{\alpha}{2}}(at)t^{1-\frac{\alpha}{2}}t^{p-1}\mathrm{d}t=\frac{2^{p-\alpha/2}\Gamma\left(\frac{1+r+p}{2}\right)}{a^{p-\alpha/2+1}\Gamma\left(\frac{1+r+\alpha-p}{2}\right)}$$

由 Mellin 反演定理,即得

$$\frac{1}{2\pi \mathrm{i}}\int_{\sigma-\mathrm{i}\infty}^{\sigma+\mathrm{i}\infty}2^{p-\alpha}\ \frac{\Gamma\left(\frac{1+r+p}{2}\right)}{\Gamma\left(\frac{1+r+\alpha-p}{2}\right)}(at)^{-p}\mathrm{d}p$$

$$=2^{-\frac{\alpha}{2}}(at)^{1-\frac{\alpha}{2}}J_{r+\frac{\alpha}{2}}(at)$$

于是便得到

$$\frac{1}{2\pi\mathrm{i}}f(x)=\frac{(2x)^{1-\frac{\alpha}{2}}}{\Gamma\left(\frac{\alpha}{2}\right)}\int_0^1\eta^{1+\frac{\alpha}{2}}J_{r+\frac{\alpha}{2}}(\eta x)\mathrm{d}\eta\cdot$$

$$\int_0^1 g(\eta\delta)\varepsilon^{r+1}(1-\delta^2)^{\frac{\alpha}{2}-1}\mathrm{d}\delta \qquad (12)$$

这个解，当 $\alpha>0$ 时是正确的，当 $\alpha\leqslant 0$ 时不适用，但可用类似方法推得 $\alpha>-2$ 时合用的解为

$$f(x)=\frac{2^{-\frac{\alpha}{2}x-\alpha}}{\Gamma\left(1+\frac{\alpha}{2}\right)}\Big[x^{1+\frac{\alpha}{2}}J_{r+\frac{\alpha}{2}}(x)$$

$$\int_0^1 y^{r+1}(1-y^2)^{\frac{\alpha}{2}}g(y)\mathrm{d}y+$$

$$\int_0^1 u^{r+1}(1-u^2)^{\frac{\alpha}{2}}\mathrm{d}u$$

$$\int_0^1 g(yu)(xy)^{2+\frac{\alpha}{2}}J_{1+r+\frac{\alpha}{2}}(xy)\mathrm{d}y\Big] \qquad (13)$$

公式(12) 对于 $\alpha>-2$ 与 $-r-1<\alpha-\frac{1}{2}<r+1$ 时有效，当 $\alpha>0$ 时可以证明解(13) 可化为解(12) 的形式，此过程较为麻烦，这里不再论述.

1994 年沈克精教授以《多重 Fourier 变换与 Hankel 变换的关系》① 为题利用多重 Fourier 变换证明 Hankel 变换的反演定理，同时，把 K 维空间的射线函数(仅依赖于到原点距离的函数) 的 Fourier 变换，归结为一维空间的 Hankel 变换，这样，由 K 元函数的 Fourier 变换成立的定理，就可推出一元函

① 摘自《安徽大学学报》(自然科学版)，1994(4)：11-15.

数的 Hankel 变换相应的定理，如下：

定义 1 设 $f(x) \in L_1(0, +\infty)$，作下面积分

$$H(p) = \int_0^{+\infty} x f(x) J_n(px) \mathrm{d}x, p > 0 \tag{1}$$

其中 $J_n(px)$ 为第一类 n 阶 Bessel 函数，(1) 的右端积分绝对收敛. $H(p)$ 称为 $f(x)$ 的 p 次 Hankel 变换.

定义 2 设 $f(x_1, x_2, \cdots, x_k)$ 是定义在 k 维空间 $\mathbf{R}_k$ 中的函数，且

$$\int_{-\infty}^{+\infty} \cdots \int_{-\infty}^{+\infty} | f(x_1, x_2, \cdots, x_k) | \mathrm{d}x_1 \mathrm{d}x_2 \cdots \mathrm{d}x_k < + \infty$$

令

$$F(s_1, s_2, \cdots, s_k) = \frac{1}{(2\pi)^{\frac{k}{2}}} \int_{-\infty}^{+\infty} \cdots \int_{-\infty}^{+\infty} f(x_1, x_2, \cdots, x_k) \mathrm{e}^{\mathrm{i}(s_1 x_1 + \cdots + s_k x_k)} \mathrm{d}x_1 \cdots \mathrm{d}x_k \tag{2}$$

则 $F(s_1, s_2, \cdots, s_k)$ 称为 $f(x_1, x_2, \cdots, x_k)$ 的 k 重 Fourier 变换.

由上述定义，可建立下面两个重要结果：

定理 1 设 $f(x) \in L_1(0, +\infty)$ 满足 Dirichlet 条件，且在 $(0, +\infty)$ 内的 Hankel 变换式由公式(1)定义，则函数 $f(x)$ 在每一连续点处均有

$$f(x) = \int_0^{+\infty} p H(p) J_n(px) \mathrm{d}p \tag{3}$$

证 由定义 2，可得二重的 Fourier 变换为

$$F(s_1, s_2) = \frac{1}{2\pi} \int_{-\infty}^{+\infty} \int_{-\infty}^{+\infty} f(x_1, x_2) \mathrm{e}^{\mathrm{i}(s_1 x_1 + s_2 x_2)} \mathrm{d}x_1 \mathrm{d}x_2 \tag{4}$$

对应的 Fourier 积分为

$$f(x_1,x_2)=\frac{1}{2\pi}\int_{-\infty}^{+\infty}\int_{-\infty}^{+\infty}F(s_1,s_2)\mathrm{e}^{-\mathrm{i}(x_1s_1+x_2s_2)}\mathrm{d}s_1\mathrm{d}s_2 \tag{5}$$

令

$$x_1=r\cos\theta,x_2=r\sin\theta,s_1=p\cos\alpha,s_2=p\sin\alpha$$

则公式(4)(5) 分别变为

$$F(p,\alpha)=\frac{1}{2\pi}\int_0^{+\infty}r\mathrm{d}r\int_0^{2\pi}f(r,\theta)\mathrm{e}^{\mathrm{i}pr\cos(\theta-\alpha)}\mathrm{d}\alpha \tag{6}$$

$$f(r,\theta)=\frac{1}{2\pi}\int_0^{+\infty}p\mathrm{d}p\int_0^{2\pi}F(p,\alpha)\mathrm{e}^{-\mathrm{i}pr\cos(\theta-\alpha)}\mathrm{d}\theta \tag{7}$$

取 $f(r,\theta)=\mathrm{e}^{-\mathrm{i}n\theta}f(r)$，于是(6) 变为

$$F(p,\alpha)=\frac{1}{2\pi}\int_0^{+\infty}rf(r)\mathrm{d}r\int_0^{2\pi}\mathrm{e}^{\mathrm{i}[pr\cos(\theta-\alpha)-n\theta]}\mathrm{d}\theta \tag{8}$$

若令 $\varphi=\alpha-\theta-\frac{\pi}{2}$，则得

$$\begin{aligned}
&\int_0^{2\pi}\mathrm{e}^{\mathrm{i}[pr\cos(\theta-\alpha)-n\theta]}\mathrm{d}\theta\\
&=-\int_{\alpha-\frac{\pi}{2}}^{\alpha-\frac{5\pi}{2}}\mathrm{e}^{\mathrm{i}\left[-pr\sin\varphi-n\left(\alpha-\frac{\pi}{2}-\varphi\right)\right]}\mathrm{d}\varphi\\
&=\mathrm{e}^{\mathrm{i}n\left(\frac{\pi}{2}-\alpha\right)}\int_{\alpha-\frac{5\pi}{2}}^{\alpha-\frac{\pi}{2}}\mathrm{e}^{\mathrm{i}(n\varphi-pr\sin\varphi)}\mathrm{d}\varphi\\
&=\mathrm{e}^{\mathrm{i}n\left(\frac{\pi}{2}-\alpha\right)}\int_{\alpha-\frac{5\pi}{2}}^{\alpha-\frac{5\pi}{2}+2\pi}\mathrm{e}^{\mathrm{i}(n\varphi-pr\sin\varphi)}\mathrm{d}\varphi\\
&=\mathrm{e}^{\mathrm{i}n\left(\frac{\pi}{2}-\alpha\right)}\int_0^{2\pi}\mathrm{e}^{\mathrm{i}(n\varphi-pr\sin\varphi)}\mathrm{d}\varphi\\
&=2\pi\mathrm{e}^{\mathrm{i}n\left(\frac{\pi}{2}-\alpha\right)}J_n(pr)
\end{aligned}$$

故得

$$\begin{aligned}
F(p,\alpha)&=\mathrm{e}^{\mathrm{i}n\left(\frac{\pi}{2}-\alpha\right)}\int_0^{+\infty}rf(r)J_n(pr)\mathrm{d}r\\
&=\mathrm{e}^{\mathrm{i}n\left(\frac{\pi}{2}-\alpha\right)}H(p)
\end{aligned}$$

于是(7)变成

$$f(r)\mathrm{e}^{-\mathrm{i}n\theta}=\frac{1}{2\pi}\int_0^{+\infty}pH(p)\mathrm{d}p\int_0^{2\pi}\mathrm{e}^{\mathrm{i}\left[n\left(\frac{\pi}{2}-\alpha\right)-pr\cos(\theta-\alpha)\right]}\mathrm{d}\alpha$$

令 $\varphi=\theta-\alpha+\frac{\pi}{2}$,可得

$$\int_0^{2\pi}\mathrm{e}^{\mathrm{i}\left[n\left(\frac{\pi}{2}-\alpha\right)-pr\cos(\theta-\alpha)\right]}\mathrm{d}\alpha=-\int_{\theta+\frac{\pi}{2}}^{\theta-\frac{3\pi}{2}}\mathrm{e}^{\mathrm{i}[n(\varphi-\theta)-pr\sin\varphi]}\mathrm{d}\varphi$$

$$=\mathrm{e}^{-\mathrm{i}n\theta}\int_0^{2\pi}\mathrm{e}^{\mathrm{i}(n\psi-pr\sin\psi)}\mathrm{d}\varphi=2\pi\mathrm{e}^{-\mathrm{i}n\theta}J_n(pr)$$

故得

$$f(r)=\int_0^{+\infty}pH(p)J_n(pr)\mathrm{d}p$$

或

$$f(x)=\int_0^{+\infty}pH(p)J_n(px)\mathrm{d}p$$

为此,得到了 Hankel 变换的反演公式.

定理 2 设 $f(x_1,x_2,\cdots,x_k)$ 是 k 维空间 $\mathbf{R}_k$ 中的可积函数, 它仅是 r 的函数, $r=\sqrt{x_1^2+x_2^2+\cdots+x_k^2}$,那么,它的 Fourier 变换 $F(s_1,s_2,\cdots,s_k)$ 就成为依赖于 p 的函数 $F(p)$, $p=\sqrt{s_1^2+s_2^2+\cdots+s_k^2}$,并且 $p^{\frac{k}{2}-1}F(p)$ 是 $r^{\frac{k}{2}-1}f(r)$ 的 $\frac{k}{2}-1$ 次 Hankel 变换.

证 $f(x_1,x_2,\cdots,x_k)$ 是 $r=\sqrt{x_1^2+x_2^2+\cdots+x_k^2}$ 的函数,由定义 2,它的 k 重 Fourier 变换为

$$F(s_1,s_2,\cdots,s_k)=\frac{1}{(2\pi)^{\frac{k}{2}}}\int_{-\infty}^{+\infty}\cdots\int_{-\infty}^{+\infty}f(r)\mathrm{e}^{\mathrm{i}(s\cdot x)}\mathrm{d}x_1\mathrm{d}x_2\cdots\mathrm{d}x_k$$

其中

$$s\cdot x=s_1x_1+s_2x_2+\cdots+s_kx_k$$

令

$$p^2 = s_1^2 + s_2^2 + \cdots + s_k^2$$

$$S_i = p\alpha_i, i = 1, 2, \cdots, k$$

作变换

$$y_1 = \sum_{i=1}^{k} \alpha_i x_i$$

$$y_j = \sum_{i=1}^{k} \alpha_{ij} x_i, j = 2, 3, \cdots, k$$

其中 α_{ij} 是正交变换的系数,故得

$$r^2 = \sum_{i=1}^{k} y_i^2$$

$$s \cdot x = \sum_{i=1}^{k} s_i x_i = p \sum_{i=1}^{k} \alpha_i x_i = p y_1$$

再令 $\lambda^2 = y_2^2 + y_3^2 + \cdots + y_k^2$,则得

$$F(s_1, s_2, \cdots, s_k)$$

$$= \frac{1}{(2\pi)^{\frac{k}{2}}} \int_{-\infty}^{+\infty} e^{ipy_1} dy_1 \int_{-\infty}^{+\infty} \cdot \cdots \cdot$$

$$\int_{-\infty}^{+\infty} f(\sqrt{y_1^2 + y_2^2 + \cdots + y_k^2} dy_2 \cdots dy_k)$$

$$= \frac{1}{(2\pi)^{\frac{k}{2}}} \int_{-\infty}^{+\infty} e^{ipy_1} dy_1 \int_{0}^{+\infty} f(\sqrt{y_1^2 + \lambda^2}) \Omega(\lambda) d\lambda \tag{9}$$

$y_2, \cdots, y_k$ 化成 $k-1$ 维极坐标的形式,其中 $\Omega(\lambda)$ 是 $k-1$ 维空间的体积元素,可写成

$$\Omega(\lambda) = \omega \lambda^{k-2} \tag{10}$$

其中 ω 是 k 的函数,与 λ 无关.

对于任意的 Φ 下式成立

$$\int_{-\infty}^{+\infty} \cdots \int_{-\infty}^{+\infty} \Phi(\sqrt{y_2^2 + \cdots + y_k^2}) dy_2 \cdots dy_k$$

$$= \int_{0}^{+\infty} \Phi(\lambda) \Omega(\lambda) d\lambda \tag{11}$$

为了决定 ω 的形式,可取特殊情形

$$\Phi(\lambda)=\mathrm{e}^{-\lambda^2}$$

(11) 的左端为

$$\int_{-\infty}^{+\infty}\cdots\int_{-\infty}^{+\infty}\mathrm{e}^{-(y_2^2+\cdots+y_k^2)}\mathrm{d}y_2\mathrm{d}y_3\cdots\mathrm{d}y_k=\left(\int_{-\infty}^{+\infty}\mathrm{e}^{-y^2}\mathrm{d}y\right)^{k-1} \tag{12}$$

(11) 的右端为

$$\omega\int_0^{+\infty}\mathrm{e}^{-\lambda^2}\lambda^{k-2}\mathrm{d}\lambda$$

由于 $\int_{-\infty}^{+\infty}\mathrm{e}^{-y^2}\mathrm{d}y=\sqrt{\pi}$,即得

$$(\sqrt{\pi})^{k-1}=\omega\ \frac{1}{2}\Gamma\left(\frac{k-1}{2}\right)$$

或

$$\omega=\frac{2\pi^{\frac{k-1}{2}}}{\Gamma\left(\frac{k-1}{2}\right)} \tag{13}$$

把(13) 与(12) 代入(9),得

$$\begin{aligned}&F(s_1,s_2,\cdots,s_k)\\&=\frac{1}{(2\pi)^{\frac{k}{2}}}\cdot\frac{2\pi^{\frac{k-1}{2}}}{\Gamma\left(\frac{k-1}{2}\right)}\int_{-\infty}^{+\infty}\mathrm{e}^{\mathrm{i}\rho y_1}\mathrm{d}y_1\int_0^{+\infty}f(\sqrt{y_1^2+\lambda^2})\lambda^{k-2}\mathrm{d}\lambda\end{aligned}$$

令 $\lambda=r\sin\varphi, y_1=r\cos\varphi$ 得

$$\begin{aligned}&F(s_1,s_2,\cdots,s_k)\\&=\frac{2\pi^{\frac{k-1}{2}}}{(2\pi)^{\frac{k}{2}}\Gamma\left(\frac{k-1}{2}\right)}\int_0^{+\infty}r^{k-1}f(r)\mathrm{d}r\int_0^{\pi}\sin^{k-2}\varphi\mathrm{e}^{\mathrm{i}\rho r\cos\varphi}\mathrm{d}\varphi\end{aligned} \tag{14}$$

由于

$$\int_0^{\pi}\sin^{k-2}\varphi e^{ix\cos\varphi}d\varphi=\frac{2^{\frac{k}{2}-1}\sqrt{\pi}\Gamma\left(\frac{k-1}{2}\right)}{x^{\frac{k}{2}-1}}J_{\frac{k-2}{2}}(x)$$

为此,(14) 变为

$$F(s_1,s_2,\cdots,s_k)=\frac{1}{p^{\frac{k}{2}-1}}\int_0^{+\infty}r(r^{\frac{k}{2}-1}f(r))J_{\frac{k-2}{2}}(pr) \tag{15}$$

(15) 右端仅是复数 p 的函数,记为

$$F(s_1,s_2,\cdots,s_k)=F(p)$$

立得

$$p^{\frac{k}{2}-1}F(p)=\int_0^{+\infty}r(r^{\frac{k}{2}-1}f(r))J_{\frac{k}{2}-1}(pr)dr$$

即 $p^{\frac{k}{2}-1}F(p)$ 是 $r^{\frac{k}{2}-1}f(r)$ 的 $\frac{k}{2}-1$ 次 Hankel 变换.

鉴此,在某些情形下,k 维空间中的 Fourier 变换式是可以化为一维空间中的 Hankel 变换式的.由此可知,根据对于多元函数的Fourier变换式所证明的任何定理,都可以推出关于 Hankel 变换式的相应定理.

特别地,依照上述方法,应用于公式

$$f(x_1,x_2,\cdots,x_k)$$
$$=\frac{1}{(2\pi)^{\frac{k}{2}}}\int_{-\infty}^{+\infty}\cdots\int_{-\infty}^{+\infty}F(s_1,s_2,\cdots,s_k)e^{-i(s\cdot x)}ds_1ds_2\cdots ds_k$$

即可推出

$$r^{\frac{k}{2}-1}f(r)=\int_0^{+\infty}p(p^{\frac{k}{2}-1}F(p))J_{\frac{k}{2}-1}(pr)dp$$

令

$$g(r)=r^{\frac{k}{2}-1}f(r),h(p)=p^{\frac{k}{2}-1}F(p),\mu=\frac{k}{2}-1$$

便可得到互逆公式

$$h(p)=\int_0^{+\infty} rg(r)J_\mu(pr)\mathrm{d}r$$

$$g(r)=\int_0^{+\infty} ph(p)J_\mu(pr)\mathrm{d}p$$

Laplace 变换是迅速发展起来的一种有效的数学方法.借助于 Laplace 变换可把微积分的运算转化为复平面的代数运算,因此,可利用它解常微分方程、偏微分方程、积分方程及差分方程,简化了求解过程,是解线性系统的重要工具,在现代自控理论中得到了广泛的应用.这些内容,在有关的教程或专著中,已屡见不鲜了.

安徽大学数学系的沈克精教授还曾给出 Laplace 变换的另一个新应用①,即利用 Laplace 变换计算广义积分,从而得到计算一类广义积分的新方法,如下:

1 Laplace 变换的概念与存在定理

定义 设函数 $f(t)$ 满足条件(C):

(i) 当 $t<0$ 时,$f(t)=0$;

(ii) 当 $t\geqslant 0$ 时,$f(t)$ 及 $f'(t)$ 除去有限个第一类间断点外而处处连续;

(iii) 当 $t\to\infty$ 时,$f(t)$ 按指数级增长,即存在两个常数 M 和 σ_0 使得 $|f(t)|\leqslant Me^{\sigma_0 t}$,$0<t<\infty$,称

$$F(p)=\int_0^{\infty} f(t)e^{-pt}\mathrm{d}t \tag{1}$$

为 $f(t)$ 的拉普拉斯变换或象函数,其中 $f(t)$ 是实变量 t 的实函数或复函数,$p=\sigma+i\omega$ 为复数,简记为

① 摘自《数学的实践与认识》,1991(1):50-55.

$$F(p)=L[f(t)]$$

而 $f(t)$ 称为 $F(p)$ 的 Laplace 反变换或原函数，简记为

$$f(t)=L^{-1}[F(p)]$$

定理 1(Laplace 变换存在定理)　设函数 $f(t)$ 满足条件(C)，则由(1) 所定义的复函数 $F(p)$ 在半平面 $\text{Re } p=\sigma>\sigma_0$ 上有定义，且是解析函数.

2　计算 $\int_0^{\infty} t^n f(t)\mathrm{d}t (n\in \mathbf{N})$ 型积分

定理 2(积分定理)　设 $f(t)$ 是可变换的，$F(p)=L[f(t)]$，则

$$L\left[\int_0^t f(t)\mathrm{d}t\right]=\frac{F(p)}{p} \tag{2}$$

定理 3(终值定理)　设 $f(t)$ 是可变换的，$F(p)=L[f(t)]$，$\lim\limits_{t\to\infty} f(t)$ 存在，且 $pF(p)$ 在 $\text{Re } p>0$ 上解析，则

$$\lim_{t\to\infty} f(t)=\lim_{p\to 0} pF(p) \tag{3}$$

由上述定理可得如下结果：

命题 1　设 $f(t)$ 是可变换的，$F(p)=L[f(t)]$ 定义在 $\text{Re } p>0$ 上，且

$$\int_0^{\infty} t^n f(t)\mathrm{d}t$$

收敛，则 $(-1)^n \lim\limits_{p\to 0} F^{(n)}(p)$ 存在，且

$$(-1)^n \lim_{p\to 0} F^{(n)}(p)=\int_0^{\infty} t^n f(t)\mathrm{d}t \tag{4}$$

证　由定理 1，$F(p)$ 在 $\text{Re } p>0$ 上解析，且由积分(1) 的一致收敛性及解析函数的无穷可微性，

可对 $F(p)$ 求 n 阶导数得

$$F^{(n)}(p)=\int_0^{\infty}(-t)^n f(t)\mathrm{e}^{-pt}\mathrm{d}t$$

即

$$L[t^n f(t)]=(-1)^n F^{(n)}(p)$$

由于 $F^{(n)}(p)$ 在 $\mathrm{Re}\,p>0$ 上可微,故 $\lim\limits_{p\to 0}(-1)^n F^{(n)}(p)$ 存在. 再由定理 2 及定理 3,有

$$L\left[\int_0^t t^n f(t)\mathrm{d}t\right]=\frac{(-1)^n F^{(n)}(p)}{p}$$

及

$$\begin{aligned}\lim_{t\to\infty}\int_0^t t^n f(t)\mathrm{d}t&=\lim_{p\to 0}p\,\frac{(-1)^n F^{(n)}(p)}{p}\\&=\lim_{p\to 0}(-1)^n F^{(n)}(p)\end{aligned}$$

即

$$\int_0^{\infty}t^n f(t)\mathrm{d}t=(-1)^n\lim_{p\to 0}F^{(n)}(p)$$

特别地,当 $n=0$ 时,有

$$\int_0^{\infty}f(t)\mathrm{d}t=\lim_{p\to 0}F(p) \tag{5}$$

例 1 计算广义积分 $\int_0^{\infty}\mathrm{e}^{-\alpha t}J_0(\omega t)\mathrm{d}t(\alpha\geqslant 0,\omega>0)$,其中 $J_0(\omega t)$ 为第一类零阶 Bessel 函数.

解 由于 $L[\mathrm{e}^{-\alpha t}J_0(\omega t)]=\dfrac{1}{\sqrt{(p+\alpha)^2+\omega^2}}$,利用(5) 可得

$$\int_0^{\infty}\mathrm{e}^{-\alpha t}J_0(\omega t)\mathrm{d}t=\lim_{p\to 0}\frac{1}{\sqrt{(p+\alpha)^2+\omega^2}}=\frac{1}{\sqrt{\alpha^2+\omega^2}}$$

当 $\alpha=0$ 时,即得 $\int_0^{\infty}J_0(\omega t)\mathrm{d}t=\dfrac{1}{\omega}$.

例 2 计算广义积分 $\int_0^{\infty}\mathrm{e}^{-\alpha t}J_r(\omega t)\mathrm{d}t(\alpha\geqslant 0,$

$\omega > 0$)，其中 $J_r(\omega t)$ 为第一类 r 阶 Bessel 函数.

解 由于 $L[J_r(\omega t)] = \dfrac{\omega^r}{\sqrt{p^2+\omega^2}(\sqrt{p^2+\omega^2}+p)^r}$，

故得

$$L[e^{-\alpha t}J_r(\omega t)]$$

$$= \frac{\omega^r}{\sqrt{(p+\alpha)^2+\omega^2}(\sqrt{(p+\alpha)^2+\omega^2}+p+\alpha)^r}$$

利用(5) 可得

$$\int_0^{\infty} e^{-\alpha t} J_r(\omega t)\mathrm{d}t$$

$$= \lim_{p\to 0} \frac{\omega^r}{\sqrt{(p+\alpha)^2+\omega^2}(\sqrt{(p+\alpha)^2+\omega^2}+p+\alpha)^r}$$

$$= \frac{\omega^r}{\sqrt{\alpha^2+\omega^2}(\sqrt{\alpha^2+\omega^2}+\alpha)^r}$$

当 $\alpha = 0$ 时，即得

$$\int_0^{\infty} J_r(\omega t)\mathrm{d}t = \frac{1}{\omega}$$

上述结果表明 $\int_0^{\infty} J_r(\omega t)\mathrm{d}t$ 的值与 r 无关.

例 3 计算广义积分 $\int_0^{\infty} t^2 e^{-\alpha t}\sin\beta t\,\mathrm{d}t$($\alpha > 0$).

解 由于 $L[e^{-\alpha t}\sin\beta t] = \dfrac{\beta}{(p+\alpha)^2+\beta^2}$

利用(4) 可得

$$\int_0^{\infty} t^2 e^{-\alpha t}\sin\beta t\,\mathrm{d}t = (-1)^2 \lim_{p\to 0}\frac{\mathrm{d}^2}{\mathrm{d}p^2}\left[\frac{\beta}{(p+\alpha)^2+\beta^2}\right]$$

$$= \lim_{p\to 0}\frac{2\beta[3(p+\alpha)^2-\beta^2]}{[(p+\alpha)^2+\beta^2]^3}$$

$$= \frac{2\beta(3\alpha^2-\beta^2)}{(\alpha^2+\beta^2)^3}$$

3　计算$\int_0^{\infty}\frac{f(t)}{t}\mathrm{d}t$型积分

定理 4(积分反演定理)　设 $f(t)$ 是可变换的，$F(p)=L[f(t)]$ 定义在 $\mathrm{Re}\ p>0$ 上，且$\int_p^{\infty}F(p)\mathrm{d}p$ 收敛，则

$$L^{-1}\left[\int_p^{\infty}F(p)\mathrm{d}p\right]=\frac{f(t)}{t} \tag{6}$$

由上述定理及命题 1，可得：

命题 2　设 $f(t)$ 是可变换的，$F(p)=L[f(t)]$ 定义在$\mathrm{Re}\ p>0$ 上，且$\int_p^{\infty}F(p)\mathrm{d}p$ 与$\int_0^{\infty}\frac{f(t)}{t}\mathrm{d}t$ 均收敛，则$\lim\limits_{p\to 0}\int_p^{\infty}F(p)\mathrm{d}p$ 存在，且

$$\int_0^{\infty}F(p)\mathrm{d}p=\int_0^{\infty}\frac{f(t)}{t}\mathrm{d}t \tag{7}$$

证　由(6) 可得

$$L\left[\frac{f(t)}{t}\right]=\int_p^{\infty}F(p)\mathrm{d}p$$

再由(5) 立得

$$\int_0^{\infty}\frac{f(t)}{t}\mathrm{d}t=\lim_{p\to 0}\int_p^{\infty}F(p)\mathrm{d}p=\int_0^{\infty}F(p)\mathrm{d}p$$

例 4　计算广义积分$\int_0^{\infty}\frac{\sin t}{t}\mathrm{d}t$.

解　由于 $L[\sin t]=\frac{1}{p^2+1}$，利用(7) 得

$$\int_0^{\infty}\frac{\sin t}{t}\mathrm{d}t=\int_0^{\infty}\frac{1}{p^2+1}\mathrm{d}p=\arctan p\Big|_0^{\infty}=\frac{\pi}{2}$$

例 5　计算广义积分$\int_0^{\infty}\frac{\cos at-\cos bt}{t}\mathrm{d}t(a>$

$0, b>0)$.

解 由于 $L[\cos at-\cos bt]=\dfrac{p}{p^2+a^2}-\dfrac{p}{p^2+b^2}$,利用(7)可得

$$\int_0^{\infty}\frac{\cos at-\cos bt}{t}\mathrm{d}t=\int_0^{\infty}\left(\frac{p}{p^2+a^2}-\frac{p}{p^2+b^2}\right)\mathrm{d}p$$

$$=\frac{1}{2}\ln\left|\frac{p^2+a^2}{p^2+b^2}\right|\Bigg|_0^{\infty}=\ln\frac{b}{a}$$

例 6 计算广义积分 $\displaystyle\int_0^{\infty}\frac{\mathrm{e}^{-at}(1-\cos bt)}{t}\mathrm{d}t$ $(a>0, b>0)$.

解 由于

$$L[\mathrm{e}^{-at}(1-\cos bt)]=\frac{1}{p+a}-\frac{p+a}{(p+a)^2+b^2}$$

利用(7)可得

$$\begin{aligned}
&\int_0^{\infty}\frac{\mathrm{e}^{-at}(1-\cos bt)}{t}\mathrm{d}t\\
&=\int_0^{\infty}\left[\frac{1}{p+a}-\frac{p+a}{(p+a)^2+b^2}\right]\mathrm{d}p\\
&=\left[\ln|p+a|-\ln\sqrt{(p+a)^2+b^2}\right]\Big|_0^{\infty}\\
&=\ln\frac{|p+a|}{\sqrt{(p+a)^2+b^2}}\Bigg|_0^{\infty}\\
&=-\ln\frac{a}{\sqrt{a^2+b^2}}\\
&=\frac{1}{2}\ln\left(1+\frac{b^2}{a^2}\right)
\end{aligned}$$

4 计算 $\int_0^{\infty}f(t,x)\mathrm{d}x\,(t>0)$ 型积分

命题 3 设 $f(t,x)$ 关于 t 是可变换的,

$F(p,x)=L[f(t,x)]$，且$\int_0^{\infty} f(t,x)\mathrm{d}x(t>0)$收敛，则

$$\int_0^{\infty} f(t,x)\mathrm{d}x = L^{-1}\left[\int_0^{\infty} F(p,x)\mathrm{d}p\right] \qquad (8)$$

证 由于对参数积分进行 Laplace 变换，在次序上是可交换的，故有

$$\begin{aligned}\int_0^{\infty} F(p,x)\mathrm{d}x &= \int_0^{\infty}\left[\int_0^{\infty} f(t,x)\mathrm{e}^{-pt}\mathrm{d}t\right]\mathrm{d}x \\ &= \int_0^{\infty}\left[\int_0^{\infty} f(t,x)\mathrm{d}x\right]\mathrm{e}^{-pt}\mathrm{d}t\end{aligned}$$

即

$$\int_0^{\infty} F(p,x)\mathrm{d}x = L\left[\int_0^{\infty} f(t,x)\mathrm{d}x\right]$$

立得

$$\int_0^{\infty} f(t,x)\mathrm{d}x = L^{-1}\left[\int_0^{\infty} F(p,x)\mathrm{d}x\right]$$

例 7 计算广义积分$\int_0^{\infty}\frac{\sin tx}{x}\mathrm{d}x(t>0)$.

解 由于

$$L\left[\frac{\sin tx}{x}\right]=\frac{1}{x}\frac{x}{p^2+x^2}=\frac{1}{p^2+x^2}=F(p,x)$$

$$\int_0^{\infty} F(p,x)\mathrm{d}x=\int_0^{\infty}\frac{1}{p^2+x^2}\mathrm{d}x=\frac{1}{p}\arctan\frac{x}{p}\Big|_0^{\infty}=\frac{\pi}{2p}$$

利用(8)可得

$$\int_0^{\infty}\frac{\sin tx}{x}\mathrm{d}x=L^{-1}\left[\frac{\pi}{2p}\right]=\frac{\pi}{2}$$

例 8 计算广义积分$\int_0^{\infty}\frac{\cos bx}{a^2+x^2}\mathrm{d}x(a>0,b>0)$.

解 选取$f(t,x)=\frac{\cos tx}{a^2+x^2}(t>0)$，由于

$$L\left[\frac{\cos tx}{a^2+x^2}\right]=\frac{p}{(a^2+x^2)(p^2+x^2)}=F(p,x)$$

$$\begin{aligned}\int_0^{\infty}F(p,x)\mathrm{d}x&=\int_0^{\infty}\frac{p}{(a^2+x^2)(p^2+x^2)}\mathrm{d}x\\&=\frac{p}{p^2-a^2}\int_0^{\infty}\left(\frac{1}{a^2+x^2}-\frac{1}{p^2+x^2}\right)\mathrm{d}x\\&=\frac{p}{p^2-a^2}\left(\frac{1}{a}\arctan\frac{x}{a}-\frac{1}{p}\arctan\frac{x}{p}\right)\Bigg|_0^{\infty}\\&=\frac{\pi}{2a(p+a)}\end{aligned}$$

利用(8) 可得

$$\int_0^{\infty}\frac{\cos tx}{a^2+x^2}\mathrm{d}x=L^{-1}\left[\frac{\pi}{2a(p+a)}\right]=\frac{\pi}{2a}\mathrm{e}^{-at}$$

取 $t=b$ 得

$$\int_0^{\infty}\frac{\cos bx}{a^2+x^2}\mathrm{d}x=\frac{\pi}{2a}\mathrm{e}^{-ab}$$

例 9　计算广义积分$\int_0^{\infty}\frac{\sin x}{x(x^2+1)}\mathrm{d}x$.

解　选取 $f(t,x)=\frac{\sin tx}{x(x^2+1)}(t>0)$，由于

$$L\left[\frac{\sin tx}{x(x^2+1)}\right]=\frac{1}{(p^2+x^2)(x^2+1)}=F(p,x)$$

$$\begin{aligned}\int_0^{\infty}F(p,x)\mathrm{d}x&=\int_0^{\infty}\frac{1}{(p^2+x^2)(x^2+1)}\mathrm{d}x\\&=\frac{\pi}{2p(p+1)}\end{aligned}$$

利用(8) 可得

$$\begin{aligned}\int_0^{\infty}\frac{\sin tx}{x(x^2+1)}\mathrm{d}x&=L^{-1}\left[\frac{\pi}{2p(p+1)}\right]\\&=\frac{\pi}{2}L^{-1}\left[\frac{1}{p}-\frac{1}{p+1}\right]\end{aligned}$$

$$=\frac{\pi}{2}(1-\mathrm{e}^{-t})$$

取 $t=1$ 得

$$\int_0^{\infty}\frac{\sin x}{x(x^2+1)}\mathrm{d}x=\frac{\pi}{2}(1-\mathrm{e}^{-1})$$

例 10 计算广义积分 $\int_0^{\infty}\frac{\sin^2 mx}{x^2}\mathrm{d}x(m>0)$.

解 选取 $f(t,x)=\frac{\sin^2 tx}{x^2}(t>0)$，由于

$$L\left[\frac{\sin^2 tx}{x^2}\right]=L\left[\frac{1-\cos 2tx}{2x^2}\right]$$

$$=\frac{1}{2x^2}\left(\frac{1}{p}-\frac{p}{p^2+4x^2}\right)=F(p,x)$$

$$\int_0^{\infty}F(p,x)\mathrm{d}x=\int_0^{\infty}\frac{2}{p(p^2+4x^2)}\mathrm{d}x$$

$$=\frac{1}{p^2}\int_0^{\infty}\frac{\mathrm{d}\left(\frac{2x}{p}\right)}{1+\left(\frac{2x}{p}\right)^2}=\frac{\pi}{2p^2}$$

利用(8) 可得

$$\int_0^{\infty}\frac{\sin^2 tx}{x^2}\mathrm{d}x=L^{-1}\left[\frac{\pi}{2p^2}\right]=\frac{\pi}{2}t$$

取 $t=m$ 得 $\int_0^{\infty}\frac{\sin^2 mx}{x^2}\mathrm{d}x=\frac{m\pi}{2}$.

2002 年内蒙古工业大学基础部的陈占华、段俊生两位教授以《关于梅林变换存在的充分条件》① 为题发表论文，对一些特殊类型的函数，给出其 Mellin 变换存在的一个充分条

① 摘自《内蒙古工业大学学报》，2002，21(1)：44-46.

件，并由此可确定出一些函数的 Mellin 变换的存在区域，如下：

1 引　言

Mellin 变换在数学、物理中起着相当重要的作用，可关于它的理论研究并不多见. 函数 $f(x)$ 的 Mellin 变换定义为

$$F(p)=M[f(x)]=\int_0^{\infty} f(x)x^{p-1}\mathrm{d}x \tag{1}$$

在一些文献①②③中，给出如下的 Mellin 变换存在的充分条件.

设 $f(x)$ 在区间 $(0,\varepsilon]$，$[E,\infty)$ 内连续，在 (ε,E) 内除有限个点外连续，$|f(x)|$ 在 (ε,E) 内可积，且

$$|f(x)|\leqslant Cx^{-a},\forall x\in(0,\varepsilon) \tag{2}$$

$$|f(x)|\leqslant Cx^{-b},\forall x\in(E,\infty) \tag{3}$$

其中 C,a,b 为常数，$a<b$，则 $f(x)$ 的 Mellin 变换在带状域 $a<\mathrm{Re}(p)<b$ 上存在且解析.

但一些非常重要的函数，如 $\cos x$，却找不到满足条件(2)(3) 的开区间 (a,b)；对于 $\sin x$，用条件(2)(3) 只能确定在区域 $-1<\mathrm{Re}(p)<0$ 上存在 Mellin 变换. 但实际上函数 $\cos x$ 存在 Mellin 变换，

---

① 马振华，刘坤林，陆璇，等. 现代应用分析卷[M]. 北京：清华大学出版社，1998.

② DAVIES B. Integral transforms and their applications[M]. New York：Springer-Verlag，1978.

③ BOCA R. Integral transforms and their applications[M]. Fla.，CRC press，1995.

$-1<\mathrm{Re}(p)<0$ 也只是 $\sin x$ 的 Mellin 变换存在域的一部分.

2 Mellin 变换存在的充分条件

对于正弦、余弦等函数,虽然当 $x\to\infty$ 时函数不趋于 0,但它们在任一有限区间上的积分是有界的.利用此特殊性,我们给出下面的对这类函数适用的判别其 Mellin 变换存在的充分条件.

定理 设函数 $f(x)$ 在 $(0,\infty)$ 内连续,且存在 $\varepsilon,E:0<\varepsilon<E<\infty$,满足

$$|f(x)|\leqslant C_1x^{-a},\ \forall x\in(0,\varepsilon),a<1,C_1\text{ 为常数} \tag{4}$$

$$\left|\int_E^A f(x)\mathrm{d}x\right|\leqslant C_2,E<A<\infty,C_2\text{ 是与 }A\text{ 无关的常数} \tag{5}$$

则在带状域 $a<\mathrm{Re}(p)<1$ 中,$f(x)$ 的 Mellin 变换 $F(p)$ 存在且解析.

证 不妨设 $0<\varepsilon<1<E,\forall M>0,\delta>0$,使 $a+\delta<1-\delta$.下面我们说明积分 $\int_0^\infty f(x)x^{p-1}\mathrm{d}x$ 在闭矩形域 $D:a+\delta\leqslant\mathrm{Re}(p)\leqslant 1-\delta,-M\leqslant\mathrm{Im}(p)\leqslant M$ 内一致收敛,将积分写为

$$\int_0^\infty f(x)x^{p-1}\mathrm{d}x$$

$$=\int_0^\varepsilon f(x)x^{p-1}\mathrm{d}x+\int_\varepsilon^E f(x)x^{p-1}\mathrm{d}x+\int_E^\infty f(x)x^{p-1}\mathrm{d}x \tag{6}$$

对于右边第一项,$|f(x)x^{p-1}|\leqslant C_1x^{-a}x^{\mathrm{Re}(p)-1}\leqslant$

$C_1x^{\delta-1}$，故积分$\int_0^{\varepsilon} f(x)x^{p-1}\mathrm{d}x$ 对 $p\in D$ 一致收敛. 对于式 (6) 右边第二项，$|f(x)x^{p-1}|=|f(x)|\cdot x^{\mathrm{Re}(p)-1}$，由于 $\varepsilon\leqslant x\leqslant E, a+\delta\leqslant\mathrm{Re}(p)\leqslant 1-\delta$，故 $x^{\mathrm{Re}(p)-1}\leqslant C$，$C$ 与 D 上 p 的位置无关，故$\int_{\varepsilon}^{E} f(x)x^{p-1}\mathrm{d}x$ 对于 $p\in D$ 一致收敛. 对于式(6) 右边第三项，$E<A_1<A_2$ 时，由分部积分公式

$$
\begin{aligned}
&\int_{A_1}^{A_2} f(x)x^{p-1}\mathrm{d}x\\
&=\int_{A_1}^{A_2}\left(\int_E^x f(t)\mathrm{d}t\right)'_x x^{p-1}\mathrm{d}x\\
&=A_2^{p-1}\int_E^{A_2} f(x)\mathrm{d}x-A_1^{p-1}\int_E^{A_1} f(x)\mathrm{d}x-\\
&\quad\int_{A_1}^{A_2}\left(\int_E^x f(t)\mathrm{d}t\right)(p-1)x^{p-2}\mathrm{d}x
\end{aligned}
$$

因而

$$
\begin{aligned}
&\left|\int_{A_1}^{A_2} f(x)x^{p-1}\mathrm{d}x\right|\\
&\leqslant C_2A_2^{\mathrm{Re}(p)-1}+C_2A_1^{\mathrm{Re}(p)-1}+\int_{A_1}^{A_2} C_2\,|p-1|\,x^{\mathrm{Re}(p)-2}\mathrm{d}x\\
&=C_2A_2^{\mathrm{Re}(p)-1}+C_2A_1^{\mathrm{Re}(p)-1}+\\
&\quad C_2\frac{|p-1|}{\mathrm{Re}(p)-1}\left(A_2^{\mathrm{Re}(p)-1}-A_1^{\mathrm{Re}(p)-1}\right)
\end{aligned}
$$

由于在闭矩形域 D 上$\dfrac{|p-1|}{\mathrm{Re}(p)-1}$ 有界，且当 A_1，$A_2\to\infty$ 时，$A_1^{\mathrm{Re}(p)-1}$ 与 $A_2^{\mathrm{Re}(p)-1}$ 均一致趋于 0，所以当 $A_1, A_2\to\infty$ 时，积分$\int_{A_1}^{A_2} f(x)x^{p-1}\mathrm{d}x$ 在 D 上一致趋于 0. 这样积分$\int_E^{\infty} f(x)x^{p-1}\mathrm{d}x$ 在闭矩形域 D 上一

致收敛. 从而积分$\int_0^{\infty} f(x)x^{p-1}\mathrm{d}x$ 在闭矩形域 D 上一致收敛.

其次对每一个 $x>0$，函数 $f(x)x^{p-1}$ 在 $a<\mathrm{Re}(p)<1$ 上解析. 因而函数 $f(x)$ 的 Mellin 变换 $F(p)=\int_0^{\infty} f(x)x^{p-1}\mathrm{d}x$ 在带状域 $a<\mathrm{Re}(p)<1$ 中存在且解析[1]. 证毕.

作为上面定理的应用，我们考察函数 $\cos\alpha x$ 和 $\sin\alpha x$ $(\alpha>0)$ 的 Mellin 变换. 易知 $M[\cos\alpha x]$ 在 $0<\mathrm{Re}(p)<1$ 上存在且解析，而 $M[\sin\alpha x]$ 在 $-1<\mathrm{Re}(p)<1$ 上存在且解析.

我们可进一步求出 $\cos\alpha x$ 和 $\sin\alpha x$ 的 Mellin 变换，考察复变函数 $f(z)=\mathrm{e}^{\mathrm{i}\alpha z}z^{p-1}(0<\mathrm{Re}(p)<1)$，取 z^{p-1} 在 $z=1$ 处取值为 1 的一支，$z=x+\mathrm{i}y=r\mathrm{e}^{\mathrm{i}\theta}$，$-\pi<\theta<\pi$. 取围道 $L=L_1+C_R+L_2+C_\varepsilon$，其中

$L_1:\varepsilon\leqslant\mathrm{Re}(z)\leqslant R,\mathrm{Im}(z)=0$，$|z|$ 增大的方向为正向

$C_R:|z|=R,0<\arg z<\dfrac{\pi}{2}$，$\arg z$ 增大的方向为正向

$L_2:\varepsilon\leqslant\mathrm{Im}(z)\leqslant R,\mathrm{Re}(z)=0$，$|z|$ 减小的方向为正向

$C_\varepsilon:|z|=\varepsilon,0<\arg z<\dfrac{\pi}{2}$，$\arg z$ 减小的方向为正向

由于 $f(z)$ 在 L 所围闭域上解析，故有$\int_L \mathrm{e}^{\mathrm{i}\alpha z}z^{p-1}\mathrm{d}z=0$，即

$$\int_\varepsilon^R \mathrm{e}^{\mathrm{i}\alpha x}x^{p-1}\mathrm{d}x+\int_{C_R}\mathrm{e}^{\mathrm{i}\alpha z}z^{p-1}\mathrm{d}z+$$

① LANG S. Complex analysis[M]. 4th ed. New York: Springer, 1999.

$$\int_R^{\varepsilon} e^{-\alpha y}(iy)^{p-1}dy+\int_{C_\varepsilon} e^{i\alpha z}z^{p-1}dz=0 \quad (7)$$

当 $R\to\infty$ 时，$z^{p-1}\to 0$ 在 C_R 上一致成立，由 Jordan 引理

$$\lim_{R\to\infty}\int_{C_R} e^{i\alpha z}z^{p-1}dz=0 \quad (8)$$

又因为当 $\varepsilon\to 0$ 时，$e^{i\alpha z}z^{p}\to 0$ 在 C_ε 上一致成立，故

$$\lim_{\varepsilon\to 0}\int_{C_\varepsilon} e^{i\alpha z}z^{p-1}dz=0 \quad (9)$$

因此，在式(7) 中令 $R\to\infty$，$\varepsilon\to 0$ 得

$$\int_0^{\infty} e^{i\alpha x}x^{p-1}dx=e^{\frac{i\pi p}{2}}\int_0^{\infty} e^{-\alpha y}y^{p-1}dy=\Gamma(p)\alpha^{-p}e^{\frac{i\pi p}{2}} \quad (10)$$

由式(10) 两边实部相等，得

$$M[\cos\alpha x]=\int_0^{\infty}(\cos\alpha x)x^{p-1}dx=\Gamma(p)\alpha^{-p}\cos\frac{\pi p}{2},0<\mathrm{Re}(p)<1 \quad (11)$$

由式(10) 两边虚部相等，并经解析延拓得

$$M[\sin\alpha x]=\int_0^{\infty}(\sin\alpha x)x^{p-1}dx=\Gamma(p)\alpha^{-p}\sin\frac{\pi p}{2},-1<\mathrm{Re}(p)<1 \quad (12)$$

2010 年湖南工程学院理学院的杨继明教授以《积分变换在一类广义积分计算中的应用》① 为题发表论文，对于一类广

① 摘自《湖南工程学院学报》，2010，20(1)：46-47.

义积分$\int_0^{+\infty}\frac{\sin x}{x}\mathrm{d}x$，为了克服利用留数定理来计算的不足，采用两类积分变换即 Fourier 变换和 Laplace 变换来计算. 通过实例计算证实了采用积分变换计算此类积分是简便、有效的，如下：

1 引 言

《高等数学(第五版)》《高等数学(第三版)》的第五章和《数学分析(第二版)》的第十章对广义积分(或称反常积分、非正常积分)进行了阐述并介绍了一些计算方法. 许多学生在学习完这些内容之后，对广义积分有了更多理解. 有些学生在阅读课外书时，发现有这样一类广义积分$\int_0^{+\infty}\frac{\sin x}{x}\mathrm{d}x$，其反常性既表现在积分区间为无穷区间，又表现为被积函数在 $x=0$ 处不连续($x=0$ 为瑕点). 比起一些文献①②③中的情形，这类广义积分当然要复杂一些，且被积函数的原函数并不好找. 在《复变函数与积分变换(第二版)》的第五章和《数学物理方法(第二版)》的第七章可以找到计算这类积分的方法，就是利用留数定理来计算，从而用复变函数的方法解

① 同济大学应用数学系. 高等数学(第五版)[M]. 北京：高等教育出版社，2002.

② 夏学文，陈世发. 高等数学(第三版)[M]. 北京：中国人民大学出版社，2009.

③ 华东师范大学数学系. 数学分析(第二版)[M]. 北京：高等教育出版社，1991.

决了某些用高等数学中的方法难以解决的积分计算问题.

首先简单回顾一下应用留数定理计算此积分的思想和步骤.

因为被积函数 $\frac{\sin x}{x}$ 是偶函数，所以 $\int_0^{+\infty}\frac{\sin x}{x}\mathrm{d}x=\frac{1}{2}\int_{-\infty}^{+\infty}\frac{\sin x}{x}\mathrm{d}x=\frac{1}{2}\mathrm{Im}\int_{-\infty}^{+\infty}\frac{\mathrm{e}^{\mathrm{i}x}}{x}\mathrm{d}x$. 在应用留数定理计算此积分时，应考虑复变积分 $\oint_C\frac{\mathrm{e}^{\mathrm{i}z}}{z}\mathrm{d}z$. 积分围道 C 由以原点为圆心、δ 为半径的小半圆弧 C_δ 和以原点为圆心、R 为半径的大半圆弧 C_R 以及直线段 $-R\rightarrow-\delta$ 和 $\delta\rightarrow R$ 构成.

$$\oint_C\frac{\mathrm{e}^{\mathrm{i}z}}{z}\mathrm{d}z$$
$$=\int_{-R}^{-\delta}\frac{\mathrm{e}^{\mathrm{i}x}}{x}\mathrm{d}x+\int_{C_\delta}\frac{\mathrm{e}^{\mathrm{i}z}}{z}\mathrm{d}z+\int_\delta^R\frac{\mathrm{e}^{\mathrm{i}x}}{x}\mathrm{d}x+\int_{C_R}\frac{\mathrm{e}^{\mathrm{i}z}}{z}\mathrm{d}z$$

在积分围道包围的区域内，因为被积函数解析，故围道积分为 0. 而根据相关引理，可知

$$\lim_{R\to\infty}\int_{C_R}\frac{\mathrm{e}^{\mathrm{i}z}}{z}\mathrm{d}z=0,\lim_{\delta\to0}\int_{C_\delta}\frac{\mathrm{e}^{\mathrm{i}z}}{z}\mathrm{d}z=-\pi\mathrm{i}$$

因此 $\int_{-\infty}^{+\infty}\frac{\mathrm{e}^{\mathrm{i}x}}{x}\mathrm{d}x=\pi\mathrm{i}$. 进而 $\int_0^{+\infty}\frac{\sin t}{t}\mathrm{d}t=\frac{\pi}{2}$.

留数定理为这类积分的计算提供了极为有效的方法. 应用留数定理计算实变函数定积分，就是把求实变函数的积分化成复变函数沿围道的积分，然后应用留数定理，使沿围道的积分计算，归结为留数计算. 要使用留数计算，需要具备两个条件：一是被积函数与某个解析函数有关；二是定积分可化为某个沿闭路的积分. 这给积分计算带来了一定的

困难.

在《复变函数与积分变换(第二版)》的第八章和第九章中分别介绍了Fourier变换和Laplace变换这两类积分变换.这些变换在求解微分方程等方面能简化计算,又具有非常特殊的物理意义,在许多领域被广泛地应用.从这两类积分变换的定义和性质可以研究发现,它们对于某些广义积分的计算,也不乏是一种简便而有效的方法和途径.

2 利用Laplace变换计算广义积分

在这种方法中,需要用到Laplace变换象函数的积分的反演性质.

定理1 设 $F(p)$ 是 $f(t)$ 的Lapace变换式 $F(p)=L[f(t)]=\int_0^{+\infty}\mathrm{e}^{-pt}f(t)\mathrm{d}t$,如果 $G(p)=\int_p^{+\infty}F(p)\mathrm{d}q$ 存在,且当 $t\to 0$ 时,$\left|\frac{f(t)}{t}\right|$ 有界,则 $\int_p^{+\infty}F(q)\mathrm{d}q$ 是 $\frac{f(t)}{t}$ 的Laplace变换式,即

$$\int_p^{+\infty}F(q)\mathrm{d}q=L\left[\frac{f(t)}{t}\right] \tag{1}$$

特别地,如果 $p\to 0$ 时,式(1)两端的积分均存在,那么有

$$\int_0^{+\infty}F(q)\mathrm{d}q=\int_0^{+\infty}\frac{f(t)}{t}\mathrm{d}t \tag{2}$$

根据以上定理,在Laplace变换及其一些性质中,如果选取 p 为某些特定的值,就可以求得一些函数的广义积分.

因为 $\sin t=\dfrac{e^{it}-e^{-it}}{2i}$，对其作 Laplace 变换，并运用 Laplace 变换的性质，得到其 Laplace 变换式为

$$\int_0^{\infty} e^{-pt}\ \frac{e^{it}-e^{-it}}{2i}dt=\frac{1}{2i}\left[\frac{1}{p-i}-\frac{1}{p+i}\right]=\frac{1}{p^2+1}$$

利用反演公式(1)，可以得到 $\dfrac{\sin t}{t}$ 的 Laplace 变换式

$$\int_p^{+\infty}\frac{1}{q^2+1}dq=\frac{\pi}{2}-\arctan p \tag{3}$$

然后令 $p\to 0$，在式(3)的两端取极限，得到

$$\int_0^{+\infty}\frac{\sin t}{t}dt=\int_0^{+\infty}\frac{1}{p^2+1}dp=\frac{\pi}{2}$$

与用留数定理计算的方法相比，这里的计算更为简便.

3　利用 Fourier 变换计算广义积分

设 $F(\omega)$ 为函数 $f(t)$ 的 Fourier 变换. 根据 Fourier 变换的定义和性质可知，通过 $f(t)$ 的 Fourier 积分表达式能得到广义积分 $\int_{-\infty}^{+\infty}F(\omega)\cos\omega t\,d\omega$ 和 $\int_{-\infty}^{+\infty}F(\omega)\sin\omega t\,d\omega$ 的值.

对于函数 $f(t)=\begin{cases}1, & |t|\leqslant\delta\\ 0, & |t|>\delta\end{cases}(\delta>0)$，作 Fourier 变换，可知它所对应的象函数为

$$\begin{aligned}F(\omega)&=\int_{-\infty}^{+\infty}f(t)e^{-i\omega t}dt=\int_{-\delta}^{\delta}e^{-i\omega t}dt\\&=\frac{1}{-i\omega}e^{-i\omega t}\Big|_{-\delta}^{\delta}\\&=2\,\frac{\sin\delta\omega}{\omega}\end{aligned}$$

通过 Fourier 逆变换可以得到 $f(t)$ 的 Fourier 积分表达式

$$
\begin{aligned}
f(t) &= \frac{1}{2\pi}\int_{-\infty}^{+\infty} \frac{2\sin \delta\omega}{\omega} \mathrm{e}^{\mathrm{i}\omega t} \mathrm{d}\omega \\
&= \frac{1}{2\pi}\int_{-\infty}^{+\infty} \frac{2\sin \delta\omega}{\omega} \cos \omega t \mathrm{d}\omega + \frac{\mathrm{i}}{2\pi}\int_{-\infty}^{+\infty} \frac{2\sin \delta\omega}{\omega} \sin \omega t \mathrm{d}\omega \\
&= \frac{2}{\pi}\int_{0}^{+\infty} \frac{\sin \delta\omega}{\omega} \cos \omega t \mathrm{d}\omega \\
&= \begin{cases} 1, & |t| < \delta \\ \dfrac{1}{2}, & |t| = \delta \\ 0, & |t| > \delta \end{cases}
\end{aligned}
$$

于是,有下列结论成立:

定理 2

$$
\int_{0}^{+\infty} \frac{\sin \delta\omega}{\omega} \cos \omega t \mathrm{d}\omega = \begin{cases} \dfrac{\pi}{2}, & |t| < \delta \\ \dfrac{\pi}{4}, & |t| = \delta \\ 0, & |t| > \delta \end{cases} \tag{4}
$$

根据以上定理,在 Laplace 变换及其一些性质中,如果选取 t 为某些特定的值,就可以求得一些函数的广义积分值.

在式(4)中令 $t=0,\delta=1$,则可得到积分 $\int_{0}^{+\infty} \frac{\sin x}{x} \mathrm{d}x = \frac{\pi}{2}$.

这种方法简单有效,能得到与用留数定理计算相同的结果.

在高等数学学习中,有些广义积分的计算过程是比较烦琐的,有些可能因为其原函数不是初等函数而无法计算.利用 Fourier 变换和 Laplace 变换的定义、性质及相关结论，2018 年佳木斯大学理学院的孙立伟、汪宏远、张志旭三位教授发表了题为《积分变换在广义积分中的应用》的论文[①]，研究了 $\int_0^{+\infty}\frac{\sin x}{x}\mathrm{d}x$，$\int_0^{+\infty}\mathrm{e}^{-x^2}\mathrm{d}x$ 等广义积分值的计算，简化了此类广义积分的计算，如下：

1　引言及预备知识

对于一些实变量的广义积分，由于其被积函数的原函数不是初等函数，在高等数学中提供的方法无法得出结果. 对于此类积分，可以通过引进参变量 t，使其成为 t 的函数，再采用积分变换的方法，并使参变量 t 取某个特殊值，以确定出其积分的值[②].

定义 1[③]　若函数 $f(t)$ 满足 Fourier 积分存在定理的条件，则在 $f(t)$ 的连续点处，有

$$f(t)=\frac{1}{2\pi}\int_{-\infty}^{+\infty}\left(\int_{-\infty}^{+\infty}f(\tau)\mathrm{e}^{-\mathrm{j}\omega\tau}\mathrm{d}\tau\right)\mathrm{e}^{\mathrm{j}\omega t}\mathrm{d}\omega \tag{1}$$

成立，设

$$F(\omega)=\int_{-\infty}^{+\infty}f(t)\mathrm{e}^{-\mathrm{j}\omega t}\mathrm{d}t \tag{2}$$

则

$$f(t)=\frac{1}{2\pi}\int_{-\infty}^{+\infty}F(\omega)\mathrm{e}^{\mathrm{j}\omega t}\mathrm{d}\omega \tag{3}$$

---

① 摘自《高师理科学刊》，2018，38(2)：19-22.

② 黄会芸. 拉普拉斯变换的应用[J]. 保山师专学报，2009，28(5)：16-18.

③ 张元林. 积分变换[M]. 4 版. 北京：高等教育出版社，2011：7-14.

称式(2)为 $f(t)$ 的 Fourier 变换,记为 $\mathscr{F}[f(t)]$,称式(3)为 $F(\omega)$ 的 Fourier 逆变换,记为 $\mathscr{F}^{-1}[F(\omega)]$.

定义 2[①] 设函数 $f(t)$ 在 $[0,+\infty)$ 上有定义,如果对复参量 $s=\beta+\mathrm{j}\omega$,积分 $F(s)=\int_0^{+\infty} f(t)\mathrm{e}^{-st}\mathrm{d}t$ 在 s 的某一域内收敛,则称 $F(s)$ 为 $f(t)$ 的 Laplace 变换,记作 $\mathscr{L}[f(t)]$,即 $F(s)=\mathscr{L}[f(t)]=\int_0^{+\infty} f(t)\mathrm{e}^{-st}\mathrm{d}t$;称 $f(t)$ 为 $F(s)$ 的 Laplace 逆变换,记为 $\mathscr{L}^{-1}[F(s)]$,即 $f(t)=\mathscr{L}^{-1}[F(s)]$.

性质 1(Fourier 变换的对称性质)[②] 若已知 $F(\omega)=\mathscr{F}[f(t)]$,则 $\mathscr{F}[F(t)]=2\pi f(-\omega)$.

性质 2(Laplace 变换中象函数的积分性质)[③] 设 $\mathscr{L}[f(t)]=F(s)$,若积分 $\int_s^{\infty}F(s)\mathrm{d}s$ 收敛,则有 $\mathscr{L}\mathscr{L}\left[\frac{f(t)}{t}\right]=\int_s^{\infty}F(s)\mathrm{d}s$.

特别地,若积分 $\int_0^{+\infty}\frac{f(t)}{t}\mathrm{d}t$ 存在,取积分下限 $s=0$,则有 $\int_0^{+\infty}\frac{f(t)}{t}\mathrm{d}t=\int_0^{+\infty}F(s)\mathrm{d}s$.

定理(瑞利定理)[④] 若记 $F(\omega)=\mathscr{F}[f(t)]$,则

---

① 冯卫国. 积分变换[M]. 2 版. 上海:上海交通大学出版社,2009:10-15.

② 杜洪艳,尤正书,侯秀梅. 复变函数与积分变换[M]. 武汉:华中师范大学出版社,2012:142-144.

③ 熊辉. 工科积分变换及其应用[M]. 北京:上海中国人民大学出版社,2011:84-94.

④ 杨绛龙,杨帆. 复变函数与积分变换[M]. 北京:科学出版社,2012:169-179.

$$\int_{-\infty}^{+\infty}[f(t)]^2\mathrm{d}t=\frac{1}{2\pi}\int_{-\infty}^{+\infty}|F(\omega)|^2\mathrm{d}\omega.$$

常见的积分变换对[①]：$\mathscr{L}[\mathrm{e}^{kt}]=\frac{1}{s-k}(\mathrm{Re}(s)>k)$；$\mathscr{L}[t^m]=\frac{m!}{s^{m+1}}(\mathrm{Re}(s)>0, m$ 为非负整数)；$\mathscr{L}[\sin(kt)]=\frac{k}{s^2+k^2}(\mathrm{Re}(s)>0)$；$\mathscr{L}[\cos(kt)]=\frac{s}{s^2+k^2}(\mathrm{Re}(s)>0)$.

2 应用实例

例 1 求$\int_0^{+\infty}\frac{\sin x}{x}\mathrm{d}x$.

由于$\frac{\sin x}{x}$的原函数不是初等函数，故在高等数学中$\int_0^{+\infty}\frac{\sin x}{x}\mathrm{d}x$ 是无法进行的. 但在积分变换中，其结果是很容易得到的.

解法 1 利用函数 $f(t)=\begin{cases}1, & |t|\leqslant 1\\ 0, & |t|>1\end{cases}$的 Fourier 积分表达式，求解$\int_0^{+\infty}\frac{\sin x}{x}\mathrm{d}x$.

根据式(1)，有

$$f(t)=\frac{1}{2\pi}\int_{-\infty}^{+\infty}\left(\int_{-\infty}^{+\infty}f(\tau)\mathrm{e}^{-\mathrm{j}\omega\tau}\mathrm{d}\tau\right)\mathrm{e}^{\mathrm{j}\omega t}\mathrm{d}\omega$$

---

① 王丽霞. 复变函数与积分变换[M]. 苏州：江苏大学出版社，2012：137-180.

$$=\frac{1}{2\pi}\int_{-\infty}^{+\infty}\left(\int_{-1}^{1}\mathrm{e}^{-\mathrm{j}\omega\tau}\mathrm{d}\tau\right)\mathrm{e}^{\mathrm{j}\omega t}\mathrm{d}\omega$$

$$=\frac{1}{2\pi}\int_{-\infty}^{+\infty}\left\{\int_{-1}^{1}[\cos(\omega\tau)-\mathrm{j}\sin(\omega\tau)]\mathrm{d}\tau\right\}\mathrm{e}^{\mathrm{j}\omega t}\mathrm{d}\omega$$

$$=\frac{1}{\pi}\int_{-\infty}^{+\infty}\frac{\sin\omega}{\omega}[\cos(\omega t)+\mathrm{j}\sin(\omega t)]\mathrm{d}\omega$$

$$=\frac{2}{\pi}\int_{0}^{+\infty}\frac{\sin\omega\cos(\omega t)}{\omega}\mathrm{d}\omega,t\neq\pm 1$$

当 $t=\pm 1$ 时，$f(t)$ 以 $\frac{f(\pm 1+0)+f(\pm 1-0)}{2}=\frac{1}{2}$ 代替.

因此 $\int_{0}^{+\infty}\frac{\sin\omega\cos(\omega t)}{\omega}\mathrm{d}\omega=\frac{\pi}{2}f(t)=\begin{cases}\frac{\pi}{2}, & |t|<1\\ \frac{\pi}{4}, & |t|=1\\ 0, & |t|>1\end{cases}$，令 $t=0$ 时，便得 $\int_{0}^{+\infty}\frac{\sin\omega}{\omega}\mathrm{d}\omega=\frac{\pi}{2}$.

解法 2 利用单个矩形脉冲函数的 Fourier 变换及 Fourier 变换的对称性质求解 $\int_{0}^{+\infty}\frac{\sin x}{x}\mathrm{d}x$.

设 $f(t)$ 为单个矩形脉冲函数，即

$$f(t)=\begin{cases}E, & |t|\leqslant 0.5\tau\\ 0, & |t|>0.5\tau\end{cases}(\tau>0)$$

易知，$f(t)$ 的 Fourier 变换为

$$F(\omega)=\mathscr{F}[f(t)]=\int_{-\infty}^{+\infty}f(t)\mathrm{e}^{-\mathrm{j}\omega t}\mathrm{d}t$$

$$=\int_{-0.5\tau}^{0.5\tau}E\mathrm{e}^{-\mathrm{j}\omega t}\mathrm{d}t=\frac{2E}{\omega}\sin\left(\frac{\omega\tau}{2}\right)$$

又因为 $f(t)$ 为偶函数，所以根据 Fourier 变换的对

称性质，可得

$$\mathscr{F}[F(t)]=\int_{-\infty}^{+\infty}\frac{2E}{t}\sin\left(\frac{\tau t}{2}\right)\mathrm{e}^{-\mathrm{j}\omega t}\,\mathrm{d}t=2\pi f(\omega)$$

即

$$2E\int_{-\infty}^{+\infty}\frac{\sin\left(\frac{\tau t}{2}\right)\cos(\omega t)}{t}\mathrm{d}t=\begin{cases}2\pi E, & |\omega|<0.5\tau\\ 0, & |\omega|>0.5\tau\end{cases}$$

令 $\omega=0,\tau=2$，得

$$4E\int_{0}^{+\infty}\frac{\sin t}{t}\mathrm{d}t=2\pi E$$

即 $\int_{0}^{+\infty}\frac{\sin t}{t}\mathrm{d}t=\frac{\pi}{2}$.

解法3[①] 利用 Laplace 变换中象函数的积分性质求解 $\int_{0}^{+\infty}\frac{\sin x}{x}\mathrm{d}x$.

令 $f(t)=\sin t$，则 $\mathscr{L}[f(t)]=\frac{1}{s^2+1}$，根据 Laplace 变换中象函数的积分性质，有 $\int_{0}^{+\infty}\frac{f(t)}{t}\mathrm{d}t=\int_{0}^{+\infty}\frac{1}{s^2+1}\mathrm{d}s=\arctan s\Big|_{0}^{+\infty}=\frac{\pi}{2}$，即 $\int_{0}^{+\infty}\frac{\sin x}{x}\mathrm{d}x=\frac{\pi}{2}$.

解法4[②] 构造函数 $f(t)=\int_{0}^{+\infty}\frac{\sin(tx)}{x}\mathrm{d}x$，利用 Laplace 逆变换求解 $\int_{0}^{+\infty}\frac{\sin x}{x}\mathrm{d}x$.

---

① 余丽琴，董玉娟. 积分变换在无穷限积分计算中的应用[J]. 教育教学论坛，2016(9)：171-173.

② 钱学明. 利用拉普拉斯变换求解几个重要的广义积分[J]. 河北北方学院学报，2008，24(3)：4-7.

对 $f(t)$ 取 Laplace 变换并交换积分顺序，可得

$$\begin{aligned}\mathscr{L}[f(t)]&=\int_0^{+\infty}\left[\int_0^{+\infty}\frac{\sin(tx)}{x}\mathrm{d}x\right]\mathrm{e}^{-st}\mathrm{d}t\\&=\int_0^{+\infty}\frac{1}{x}\left[\int_0^{+\infty}\sin(tx)\mathrm{e}^{-st}\mathrm{d}t\right]\mathrm{d}x\\&=\int_0^{+\infty}\frac{1}{x}\frac{x}{s^2+x^2}\mathrm{d}x\\&=\frac{1}{s}\arctan\frac{x}{s}\Big|_0^{+\infty}=\frac{1}{s}\frac{\pi}{2}\end{aligned}$$

取 Laplace 逆变换，有 $f(t)=\int_0^{+\infty}\frac{\sin(tx)}{x}\mathrm{d}x=\mathscr{L}^{-1}\left[\frac{1}{s}\frac{\pi}{2}\right]=\frac{\pi}{2}\mathscr{L}^{-1}\left[\frac{1}{s}\right]=\frac{\pi}{2}$，取 $t=1$，即得 $\int_0^{+\infty}\frac{\sin x}{x}\mathrm{d}x=\frac{\pi}{2}$.

解法 5①②③ 利用瑞利定理求解 $\int_0^{+\infty}\frac{\sin x}{x}\mathrm{d}x=\frac{\pi}{2}$.

根据式(2) 可求得单个矩形脉冲函数 $f(t)=\begin{cases}0.5, & |t|<1\\0, & |t|>1\end{cases}$ 的 Fourier 变换为 $\mathscr{F}[f(t)]=\frac{\sin\omega}{\omega}=F(\omega)$. 由瑞利定理可知

$$\int_{-\infty}^{+\infty}\frac{\sin^2 t}{t^2}\mathrm{d}t=\int_{-\infty}^{+\infty}\frac{\sin^2\omega}{\omega^2}\mathrm{d}\omega$$

---

① 别荣军，谢娟，刘华勇. 拉普拉斯变换在高等数学中的应用尝试[J]. 广东技术师范学院学报，2013(12)：125-128.

② 王文平. 应用 Laplace 变换计算两类广义积分[J]. 武汉船舶职业技术学院学报，2014(5)：65-66.

③ 林敏. 利用积分变换求实积分[J]. 文理导航，2017(278)：28.

$$=2\pi\int_{-\infty}^{+\infty}|f(t)|^2\mathrm{d}t$$

$$=2\pi\int_{-1}^{1}\left(\frac{1}{2}\right)^2\mathrm{d}t=\pi$$

因为

$$\int_{-\infty}^{+\infty}\frac{\sin^2 t}{t^2}\mathrm{d}t=2\int_0^{+\infty}\frac{\sin^2 t}{t^2}\mathrm{d}t=2\int_0^{+\infty}\sin^2 t\mathrm{d}\left(-\frac{1}{t}\right)$$

$$=-2\frac{1}{t}\sin^2 t\Big|_0^{+\infty}+2\int_0^{+\infty}\frac{\sin(2t)}{t}\mathrm{d}t$$

$$=2\int_0^{+\infty}\frac{\sin(2t)}{2t}\mathrm{d}(2t)$$

所以 $2\int_0^{+\infty}\frac{\sin(2t)}{2t}\mathrm{d}(2t)=\pi$，取 $x=2t$，则得 $\int_0^{+\infty}\frac{\sin x}{x}\mathrm{d}x=\frac{\pi}{2}$.

例 2 求$\int_0^{+\infty}\mathrm{e}^{-x^2}\mathrm{d}x$.

在高等数学中，积分$\int_0^{+\infty}\mathrm{e}^{-x^2}\mathrm{d}x$ 不能用初等函数表示，故不能直接进行计算. $\int_0^{+\infty}\mathrm{e}^{-x^2}\mathrm{d}x$ 的计算需要利用二重积分$\iint_D\mathrm{e}^{-x^2-y^2}\mathrm{d}x\mathrm{d}y$ 的结果，并结合极限的夹逼准则才能够得出，计算过程比较烦琐. 可以通过积分变换的方法得到$\int_0^{+\infty}\mathrm{e}^{-x^2}\mathrm{d}x$.

解法 1 钟形脉冲函数 $f(t)=A\mathrm{e}^{-\beta t^2}\ (\beta>0)$ 是工程中常见的一个函数，根据 Fourier 变换的定义，可得

$$f(t)=\frac{1}{2\pi}\int_{-\infty}^{+\infty}F(\omega)\mathrm{e}^{\mathrm{j}\omega t}\mathrm{d}\omega=\frac{A}{\sqrt{\pi\beta}}\int_0^{+\infty}\mathrm{e}^{-\frac{\omega^2}{4\beta}}\cos(\omega t)\mathrm{d}\omega$$

整理得

$$\int_0^{+\infty} \mathrm{e}^{-\frac{\omega^2}{4\beta}} \cos(\omega t)\mathrm{d}\omega = \frac{\sqrt{\pi\beta}}{A} f(t) = \sqrt{\pi\beta}\,\mathrm{e}^{-\beta t^2}$$

取 $t=0,\beta=\frac{1}{4}$，可得$\int_0^{+\infty} \mathrm{e}^{-x^2}\mathrm{d}x = \frac{\sqrt{\pi}}{2}$.

解法 2　对函数 $f(t)=\int_0^{+\infty} \mathrm{e}^{-tx^2}\mathrm{d}x$ 取 Laplace 变换，可得

$$\begin{aligned}\mathscr{L}[f(t)] &= \int_0^{+\infty} \left(\int_0^{+\infty} \mathrm{e}^{-tx^2}\mathrm{d}x\right)\mathrm{e}^{-st}\mathrm{d}t \\ &= \int_0^{+\infty} \mathrm{d}x \int_0^{+\infty} \mathrm{e}^{-x^2 t}\mathrm{e}^{-st}\mathrm{d}t \\ &= \int_0^{+\infty} \frac{1}{s+x^2}\mathrm{d}x \\ &= \frac{1}{\sqrt{s}}\arctan\left(\frac{x}{\sqrt{s}}\right)\Big|_0^{+\infty} \\ &= \frac{1}{\sqrt{s}}\,\frac{\pi}{2}\end{aligned}$$

于是

$$\begin{aligned}f(t) = \int_0^{+\infty} \mathrm{e}^{-tx^2}\mathrm{d}x &= \mathscr{L}^{-1}\left[\frac{1}{\sqrt{s}}\,\frac{\pi}{2}\right] \\ &= \frac{\pi}{2}\mathscr{L}^{-1}\left[\frac{1}{\sqrt{s}}\right]\end{aligned}$$

由于 $\mathscr{L}[t^{-0.5}]=\frac{\sqrt{\pi}}{\sqrt{s}}$，因此$\int_0^{+\infty} \mathrm{e}^{-tx^2}\mathrm{d}x = \frac{\pi}{2}\,\frac{t^{-\frac{1}{2}}}{\sqrt{\pi}}$.

取 $t=1$，则得$\int_0^{+\infty} \mathrm{e}^{-x^2}\mathrm{d}x = \frac{\sqrt{\pi}}{2}$.

在高等数学中，对于$\int_0^{+\infty} x^m \mathrm{e}^{-nx}\mathrm{d}x\,(m\in\mathbf{N},n>0)$，需要反复利用分部积分公式建立递推公式才能

解决，烦琐程度不言而喻. 若利用常见变换对 $\mathscr{L}[t^m]=\frac{m!}{s^{m+1}}(\mathrm{Re}(s)>0, m$ 为非负整数)，可直接得出积分值，即 $\int_0^{+\infty} x^m e^{-nx}\,dx=\frac{m!}{n^{m+1}}(m\in\mathbf{N}, n>0)$，有了此结论，计算该种类型的积分就变得十分简单.

类似地，对 $\int_0^{+\infty} e^{-pt}\sin\omega t\,dt$ 和 $\int_0^{+\infty} e^{-pt}\sin\omega t\,dt$ 的计算，可以利用常见变换对 $\mathscr{L}[\sin(kt)]=\frac{k}{s^2+k^2}(\mathrm{Re}(s)>0)$ 和 $\mathscr{L}[\cos(kt)]=\frac{s}{s^2+k^2}(\mathrm{Re}(s)>0)$ 得出结果，即 $\int_0^{+\infty} e^{-pt}\sin(\omega t)dt=\frac{\omega}{p^2+\omega^2}(\omega, p\in\mathbf{R}^+)$，$\int_0^{+\infty} e^{-pt}\cos(\omega t)dt=\frac{p}{p^2+\omega^2}(\omega, p\in\mathbf{R}^+)$.

内蒙古大学硕士巴森在其导师乌云高娃教授的指导下，2019 年 5 月完成了题为《广义超几何级数及其若干应用》的学位论文，在其引言部分指出：

> 组合数学(Com binatorics) 是纯数学当中的一个重要组成部分，它主要是研究离散、有限或可数的数学结构. 普遍认为组合数学最早起源于古老的幻方问题，在公元前 2200 年中国人发现了 3 阶神农幻方，直到 15 世纪欧洲人发现了 4 阶幻方. 而中国最早的组合数学理论源于宋朝时期的“贾宪三角”，后来杨辉将其完善，所以我们普遍称之为“杨辉三角”，在西方直到 1654 年由 Pascal 提出，比中国晚了 400 多年. 关于组合数学的第一部著作是《论组合的

艺术》,它是由德国数学家Leibniz在1666年所完成的,书中第一次运用了组合论一词. 时间来到18世纪,著名数学家Euler对组合数学做出了巨大贡献,并且成功运用组合学理论解决了著名的哥尼斯堡七桥问题. 在接下来的两个多世纪里组合数学的发展与应用得以充分体现.

在美国数学学会的学科分类中,组合数学分为以下不同的五个学科,它们分别是计数组合数、设计理论、图论、极值组合、代数组合. 其中计数组合学当中又包含超几何级数、发生函数、整数分拆、反演原理、Polya计数定理,等等. 而计数组合学当中有一类重要的分支,那就是超几何级数. 下文着重介绍超几何级数的发展理论. 由于超几何级数的求和公式和变换公式在计数组合学这一范围内有着极其重要的作用,因而寻找和证明超几何级数求和公式和变换公式一直都是数学家感兴趣的问题. 寻找超几何级数的求和公式和变换公式的方法有很多种,例如:传统的变换方法、计算机代数的WZ方法以及组合反演方法,等等.

近年来,越来越多的学者热衷于对超几何级数的研究,尤其是对超几何级数的求和公式和变换公式的研究. 在1655年,著名学者John Wallis在*Arithmetica Infinitorum*一书中首次提出了超几何(hypergeometric)这一术语. 在接下来的时间里越来越多的组合数学家和学者开始研究超几何级数的较为简单情形. 简单的超几何级数的一般形式如下

$$
{}_2F_1[a,b;c;z]={}_2F_1\begin{bmatrix}a,b\\ & ;z\\ c\end{bmatrix}=\sum_{k=0}^{\infty}\frac{(a)_k(b)_k}{(c)_k}\frac{z^k}{k!}
$$

称为 Gauss 级数①. 1748 年数学家 Euler 给出 Gauss 级数的许多结果②,例如,著名的关系式

$$
{}_2F_1[-n,b;c;1]=(1-z)^{c+n-b}\ {}_2F_1[c+n,c-b;c;z]
$$

1770 年数学家 Vandermonde 利用超几何级数给出了二项式定理的推广形式

$$
{}_2F_1[-n,b;c;z]=\frac{(c-n_n)}{(c)_n}
$$

1812 年数学家 Gauss 在他的博士论文 *Disquisitiones generales circa seriem infinitam*③ 中给出了著名的 Gauss 求和定理

$$
{}_2F_1[a,b;c;1]=\frac{\Gamma(c)\Gamma(c-a-b)}{\Gamma(c-a)\Gamma(c-b)},\Re(c-a-b)>0
$$

并讨论其收敛性条件. 1836 年数学家 Kummer 证明了 Gauss 级数是二阶微分方程④⑤⑥

---

① ANDREWS G E, ASKEY R, ROY R. Special functions[M]. Cambridge university press, 2000.

② EULER L. Introduction in analysin infinitorum[M]. MM Bousquet, 1748.

③ GAUSS G F. Disquisitiones generales circa seriem infinitam[J]. Soc. Reg. Sci. 1813, 2: 1870-1933.

④ BATEMAN H. Higher transcendental functions[M]. New York: McGraw-Hill, 1953.

⑤ KUMMER E E. über die hypergeometrische Reihe[J]. Journal für die reine und angewandte Mathematik, 1836, 15: 39-83.

⑥ SRIVASTAVA H M, KARLSSON P W. Multiple Gaussian hypergeometric series[M]. Ellis Horwood, 1985.

$$z(1-z)\frac{\mathrm{d}^2 y}{\mathrm{d}z^2}+\{c-(a+b+1)z\}\frac{\mathrm{d}y}{\mathrm{d}z}-aby=0$$

的解,同时 Kummer 还给出了 24 个超几何级数类型的解,后来也称为 Kummer 24 解.使用积分方法表示 Gauss 级数应当追溯到 Euler,Euler 是第一个应用积分方法将 Gauss 级数表示出来的①

$$\begin{aligned}&{}_2F_1[a,b;c;z]\\&=\frac{\Gamma(c)}{\Gamma(b)\Gamma(c-b)}\int_0^1 t^{b-1}(1-t)^{c-b-1}(1-zt)^{-a}\mathrm{d}t\end{aligned}$$

1908 年数学家 Barnes 使用了积分围道表示出 Gauss 级数,也就是著名的 Barnes 定理②

$$\begin{aligned}&\frac{\Gamma(a)}{\Gamma(b)\Gamma(c)}\,{}_2F_1[a,b;c;z]\\&=\frac{1}{2\pi}\int_{-\mathrm{i}\infty}^{\mathrm{i}\infty}\frac{\Gamma(t)\Gamma(a-t)\Gamma(b-t)}{\Gamma(c-t)}(-z)^{-t}\mathrm{d}t\end{aligned}$$

在接下来的一段时间里,一些数学家对 Gauss 级数进行推广,Gauss 级数推广方法有很多种. Clausen③ 首次对 Gauss 级数增加了参数进而推广

---

① EULER L. Institutiones calculi integralis[M]. Academia Imperialis Scientiarum, 1792.

② BARNES E W. A new development of the theory of the hypergeometric functions[J]. Proceedings of the London Mathematical Society, 1908, 2(1): 141-177.

③ CLAUSEN T. Ueber die Fälle wenn die Reihe $y=1+\frac{\alpha\cdot\beta}{1\cdot\gamma}x+\frac{\alpha(\alpha+1)\cdot\beta(\beta+1)}{1\cdot 2\cdot\gamma(\gamma_1)}x^2+etc.$ ein quadrat von der Form $y=1+\frac{\alpha\cdot\beta\cdot\delta}{1\cdot\gamma\cdot\epsilon}x+\frac{\alpha(\alpha+1)\cdot\beta(\beta+1)\cdot\delta(\delta+1)}{1\cdot 2\cdot\gamma(\gamma_1)\cdot\epsilon(\epsilon+1)}x^2+etc$ hat[J]. für Math, 1828, 3: 89-95.

了 Gauss 级数的概念. 接下来 Saalschütz①, Dixon②和 Dougall③ 等给出了一般形式超几何级数的许多求和公式. Bailey 和 Whipple 对超几何级数进行了系统的阐述和研究，并形成了完善的理论体系. 在 1935 年 Bailey④ 出版的 *Generalized Hypergeometric Series* 一书和在 1966 年 Slater⑤ 出版的 *Generalized Hypergeometric Functions* 一书对一般形式的超几何级数进行了详细的讨论. 正是由于越来越多的学者热衷于超几何级数的研究，才有了超几何级数的蓬勃发展.

目前，研究和证明超几何级数恒等式主要有以下三种方法：传统的变换方法、计算机代数的 WZ 方法、组合反演⑥⑦⑧. 变换方法是由简单的组合恒等式来推导出较为复杂的恒等式，其主要思想是使用 Bailey 变换. 在 1951—1952 年间，Slater 应用了 Bailey 变换得到 130 多个 Rogers-Ramanujan 类型的恒等式. 计算机代数的 WZ 方法最早是由数学家

---

① SAALSCHÜTZ L. Eine summationsformel[J]. Zeitschr. Math. Phys, 1890, 35: 186-188.

② DIXON A C. Summation of a certain series[J]. Proceedings of the London Mathematical Society, 1902, 1(1): 284-291.

③ DOUGALL J. On Vandermonde's theorem, and some more general expansions[J]. Proceedings of the Edinburgh Mathematical Society, 1906, 25: 114-132.

④ BAILEY W N. Generalized hypergeometric series[M]. Cambridge University Press, 1935.

⑤ SLATER L J. Generalized hypergeometric functions[M]. Cambridge University Press, 1966.

⑥ 刘俊同. 基本超几何级数及其应用[D]. 大连理工大学, 2006.

⑦ 王晓霞. Abel 分部求和法与经典超几何级数[D]. 大连理工大学, 2007.

⑧ 魏传安. 反演技巧在组合恒等式中的应用[D]. 大连理工大学, 2006.

H. Wilf 和 D. Zeilberger 提出，该方法是证明组合恒等式的机械化方法，该方法本身易懂而且对绝大多数的等式是有效的. 因为它用到 Gosper 算法，所以应用范围受 Gosper 算法的限制，并应用 Gosper 算法实现了对 Gosper 可和的超几何级数有效. 组合反演方法就是利用反演关系发现和证明恒等式. 下文将重点介绍一下组合反演方法在超几何级数中的应用.

众所周知，在组合数学中有许多的反演公式，例如：经典 Möbius 反演、Lagrange 反演、Stirling 反演，等等，其中有一类重要的反演就是矩阵反演. 矩阵反演是指一对序列或者是级数相互可以表示的互反关系，其一般形式如下：对于任意非负整数 n，序列 $f(n)$ 和 $g(n)$ 满足如下关系式①

$$\begin{cases} f(n) = \sum_{k=0}^{n} C_{n,k} g(k) \\ g(n) = \sum_{k=0}^{n} D_{n,k} f(k) \end{cases}$$

若其中一个成立，则另一个等式也成立等价于系数矩阵 $\boldsymbol{C} = (c_{ij})$，$\boldsymbol{D} = (d_{ij})$ 互逆. 因此只需构造两个互逆的上三角矩阵即可得到一对互反公式.

正是由于矩阵反演关系在组合数学与特殊函数论中有着极其重要的作用，因此被广泛地应用于超几何级数恒等式的推导. 在 20 世纪 60 年代 Riordan 第一次系统地给出了许多矩阵反演公式. 1974 年 Gould 和徐利治给出了著名的 Gould-hsu 反

① 陈晓静. 组合反演关系的若干问题的研究[D]. 苏州大学，2007.

演公式①,其形式如下

$$\begin{cases} f(n) = \sum_{k=0}^{n} (-1)^k \binom{n}{k} \psi(k,n) g(k) \\ g(n) = \sum_{k=0}^{n} (-1)^k \binom{n}{k} (a_{k+1} + kb_{k+1}) \psi(n,k+1)^{-1} f(k) \end{cases}$$

其中$\{a_k\}$和$\{b_k\}$是任意选的实数和复数序列,且

$$\psi(x,0) \equiv 1, \psi(x,n) = \prod_{i=1}^{n} (a_i + xb_i), n = 1,2,\cdots.$$

许多反演都是 Gould-hsu 反演公式的特例,例如:Legendre反演、Chebishev反演、二项式反演,等等.此公式在学术界引起广泛的应用. Gould-hsu 反演还有分段形式的推广,其一般形式如下

$$\Omega(n) = \sum_{k=0}^{\infty} \binom{n}{2k} \frac{c_k + 2kd_k}{\phi(n,k)\psi(n,k+1)} f(k) - \sum_{k=0}^{\infty} \binom{n}{2k+1} \frac{a_k + (2k+1)b_k}{\phi(n,k+1)\psi(n,k+1)} g(k)$$

等价于方程组

$$\begin{cases} f(n) = \sum_{k=0}^{2n} (-1)^k \binom{2n}{k} \phi(k,n) \psi(k,n) \Omega(k) \\ g(n) = \sum_{k=0}^{n} (-1)^k \binom{2n+1}{k} \phi(k,n) \psi(k,n+1) \Omega(k) \end{cases}$$

其中$\{a_k\}$,$\{b_k\}$,$\{c_k\}$,$\{d_k\}$是复数序列,且$\phi(x,0) \equiv 1$,

$$\phi(x,m) = \prod_{i=0}^{m-1} (a_i + xb_i), m = 1,2,\cdots, \psi(x,0) \equiv 1,$$

① GOULD H W, HSU L C. Some new inverse series relations[J]. Duke Mathematical Journal, 1973, 40(4): 885-891.

$\psi(x,n)=\prod_{i=1}^{n-1}(a_i+xb_i)$，$n=1,2,\cdots$. 在 2002 年初文昌应用 Gould 是由 Carlitz 给出，因而 Gould-hsu 反演还有 q- 模拟形式的推广也称为 Carlitz 反演[①]，其一般形式如下

$$\begin{cases} f(n)=\sum_{k=0}^{n}(-1)^k q^{\binom{k}{2}}\begin{bmatrix} n \\ k \end{bmatrix}_q \psi(k,n,q)g(k) \\ g(n)=\sum_{k=0}^{n}(-1)^k q^{\binom{k+1}{2}-kn}\begin{bmatrix} n \\ k \end{bmatrix}_q (a_{k+1}+q^{-k}b_{k+1})\psi(n,k+1,q)^{-1}f(k) \end{cases}$$

Carlitz 反演被广泛应用于基本超几何级数恒等式.

Laplace 变换作为积分变换的重要内容，在通信类、控制类、电气类等专业课中有着广泛的运用[②③④]. 如控制工程中研究阻尼振动需要用到 Dirichlet 积分，工程热物理中研究热传导需要用到 Poisson 积分，理论光学中研究光的衍射需要用到菲涅尔积分等. 高等数学运用传统的积分方法求解这类积分显得非常复杂，在运算中还需要特殊的运算技巧[⑤⑥⑦]. 物理学与工程(电路、线性复杂网络等) 方面的许多问题都可

---

① CARLITZ L. Some inverse relations[J]. Duke Mathematical Journal, 1973, 40(4): 893-901.

② 管致中，夏恭恪，孟桥. 信号与线性系统[M]. 北京：高等教育出版社，2016.

③ 滕岩梅. 积分变换中常见问题[J]. 大学数学，2015，31(1)：105-109.

④ 周从会. 应用拉普拉斯变换解决高等数学中的一些问题[J]. 高等数学研究，2016，19(4)：61-62.

⑤ 李景和，金少华. 拉普拉斯变换微分性质的若干应用[J]. 高等数学研究，2015，18(1)：76-78.

⑥ 陈荣军，文传军. 复变函数与积分变换[M]. 南京：南京大学出版社，2015.

⑦ 胡政发. 复变函数与积分变换[M]. 上海：同济大学出版社，2015.

以归结为微分方程的定解来考虑①②③④⑤，通过 Laplace 变换可以很方便地对微分方程进行求解. 2019 年安徽科技学院信息与网络工程学院的董姗姗，蚌埠学院计算机工程学院的宫原野两位教授以《拉普拉斯变换在广义积分及微分方程求解中的应用》⑥ 为题发表论文，通过大量的例子，重点讨论 Laplace 变换在广义积分与微分方程中的应用，如下：

1 预备知识

定义 设函数 $f(t)$ 在区间 $(0, +\infty)$ 内有定义，且积分 $\int_0^{+\infty} f(t)\mathrm{e}^{-st}\,\mathrm{d}t$ 在 $\mathrm{Re}\, s > c$ 时收敛，由此积分结果记为 $F(s)$，称为函数 $f(t)$ 的 Laplace 变换，记为 $F(s) = L\{f(t)\}$，即

$$L\{f(t)\} = F(s) = \int_0^{+\infty} f(t)\mathrm{e}^{-st}\,\mathrm{d}t \tag{1}$$

式中，$F(s)$ 称为 $f(t)$ 的象函数，$f(t)$ 称为 $F(s)$ 的原函数.

根据式(1) 可以得到如下结论：若象函数 $F(s)$

---

① 田垒. 拉普拉斯变换在微分方程中的应用[D]. 安庆：安庆师范大学，2016.

② 高伟航，宫成春，王鹏鲲. 高阶常微分方程的拉普拉斯变换新解[J]. 高等数学研究，2018，21(1)：100-103.

③ 李高翔. 拉普拉斯变换在微分方程组求解中的应用[J]. 高等继续教育学报，2009(3)：22-24.

④ 田慧竹，宋从芝. 利用拉普拉斯变换求解微分方程[J]. 高等数学研究，2012，15(1)：67-69.

⑤ 李连忠，何乐亮. 拉普拉斯变换应用的一个推广[J]. 山东师范大学学报(自然科学版)，2007，22(1)：148-151.

⑥ 摘自《江汉大学学报(自然科学版)》，2019，47(3)：227-230.

的所有奇点都位于 y 轴左侧，则

$$\int_0^{+\infty} f(t)\,\mathrm{d}t = \lim_{s \to 0} F(s) \tag{2}$$

性质 1 设 $L\{f(t)\} = F(s)$，若函数 $f(t)$ 可导，则

$$L\{f^{(n)}(t)\} = s^n F(s) - s^{n-1} f(0) - s^{n-2} f'(0) - \cdots - f^{(n-1)}(0) \tag{3}$$

性质 2 设 $L\{f(t)\} = F(s)$，若函数 $f(t)$ 可积，则

$$L\left\{\int_0^t f(t)\,\mathrm{d}t\right\} = \frac{F(s)}{s} \tag{4}$$

性质 3 设 $L\{f(t)\} = F(s)$，若象函数 $F(s)$ 可导，则

$$L\{t^n f(t)\} = (-1)^n F^{(n)}(s) \tag{5}$$

性质 4 设 $L\{f(t)\} = F(s)$，若象函数 $F(s)$ 可积，则

$$L\left\{\frac{f(t)}{t}\right\} = \int_s^{\infty} F(s)\,\mathrm{d}s \tag{6}$$

性质 5 设 $L\{f(t)\} = F(s)$，则

$$L\{f(t)\mathrm{e}^{at}\} = F(s-a) \tag{7}$$

2　利用 Laplace 变换求解广义积分

例 1 计算 $\int_0^{+\infty} \frac{\cos ax - \cos bx}{x}\,\mathrm{d}x, a > 0, b > 0.$

解 上述积分为傅汝兰尼积分的一个推广，利用 Laplace 变换求解. 由

$$L[\cos ax] = \frac{s}{s^2 + a^2}, L[\cos bx] = \frac{s}{s^2 + b^2}$$

则

$$L[\cos ax - \cos bx] = \frac{s}{s^2 + a^2} - \frac{s}{s^2 + b^2}$$

根据式(6)可知

$$L\left[\frac{\cos ax - \cos bx}{x}\right] = \int_0^{+\infty} \frac{\cos ax - \cos bx}{x} e^{-sx} dx$$
$$= \int_s^{+\infty} \left(\frac{s}{s^2 + a^2} - \frac{s}{s^2 + b^2}\right) ds$$
$$= -\frac{1}{2} \ln \frac{s^2 + a^2}{s^2 + b^2}$$

再根据式(2)可知

$$\int_0^{+\infty} \frac{\cos ax - \cos bx}{x} dx = \lim_{s \to 0} F(s) = \ln \frac{b}{a}$$

例 2 计算 Dirichlet 积分$\int_0^{+\infty} \frac{\sin x}{x} dx$.

解法 1 由于 Dirichlet 积分收敛,引入参变量 t,使其成为 t 的函数. 令 $f(t) = \int_0^{+\infty} \frac{\sin xt}{x} dx$,设 $L[f(t)] = F(s)$,则有

$$F(s) = \int_0^{+\infty} \int_0^{+\infty} \frac{\sin xt}{x} e^{-st} dt dx$$
$$= \int_0^{+\infty} \frac{1}{x} \left[\int_0^{+\infty} \sin xt e^{-st} dt\right] dx$$
$$= \int_0^{+\infty} \frac{1}{x} \cdot \frac{x}{s^2 + x^2} dx$$
$$= \frac{1}{s} \int_0^{+\infty} \frac{1}{\left(\frac{x}{s}\right)^2 + 1} d\left(\frac{x}{s}\right)$$
$$= \frac{1}{s} \cdot \frac{\pi}{2}$$

再取 $F(s)$ 的 Laplace 反变换,则 $f(t) = L^{-1}\{F(s)\} = \frac{\pi}{2} \varepsilon(t)$. 取 $t = 1$,即$\int_0^{+\infty} \frac{\sin x}{x} dx = f(1) = \frac{\pi}{2}$.

解法 2 根据 $L[\sin x]=\dfrac{1}{x^2+1}$，再根据其积分性质可知

$$L\left[\frac{\sin x}{x}\right]=\int_x^{+\infty}\frac{1}{x^2+1}\mathrm{d}x=\arctan x\,\Big|_x^{+\infty}$$

$$=\frac{\pi}{2}-\arctan x$$

再根据式(2)可得

$$\int_0^{+\infty}\frac{\sin x}{x}\mathrm{d}x=\lim_{x\to 0}\left(\frac{\pi}{2}-\arctan x\right)=\frac{\pi}{2}$$

解法 3 根据式(6)可知

$$\int_0^{+\infty}\frac{\sin x}{x}\mathrm{d}x=\int_0^{+\infty}L[\sin x]\mathrm{d}s$$

$$=\int_0^{+\infty}\frac{1}{s^2+1}\mathrm{d}s$$

$$=\arctan s\,\Big|_0^{+\infty}$$

$$=\frac{\pi}{2}$$

例 3 计算 $\int_0^{+\infty}\dfrac{\sin x}{x}\mathrm{e}^{-ax}\,\mathrm{d}x, a>0$.

解 上述积分为 Dirichlet 积分的推广，由 $L[\sin x]=\dfrac{1}{x^2+1}$，根据式(7)可得

$$L[\sin x\cdot \mathrm{e}^{-ax}]=\frac{1}{(s+a)^2+1}$$

根据象函数的积分性质可知

$$L\left[\frac{\sin x}{x}\mathrm{e}^{-ax}\right]$$

$$=\int_s^{+\infty}\frac{1}{(s+a)^2+1}\mathrm{d}s$$

$$=\arctan(s+a)\,\Big|_s^{+\infty}$$

$$=\frac{\pi}{2}-\arctan(s+a)$$

根据式(2) 可知

$$\int_0^{+\infty}\frac{\sin x}{x}e^{-ax}dx$$

$$=\lim_{s\to 0}F(s)$$

$$=\lim_{s\to 0}\left(\frac{\pi}{2}-\arctan(s+a)\right)$$

$$=\frac{\pi}{2}-\arctan a$$

例 4 计算$\int_0^{+\infty}xe^{-sx}\sin x\,dx$.

解 由积分 $L[\sin\ x]=\int_0^{+\infty}\sin\ e^{-sx}dx=\frac{1}{s^2+1}$,根据式(4) 可知

$$L[x\sin x]=\int_0^{+\infty}xe^{-sx}\sin x\,dx$$

$$=-\left(\frac{1}{s^2+1}\right)'=\frac{2s}{(s^2+1)^2}$$

例 5 计算 Euler-Poisson 积分$\int_0^{+\infty}e^{-x^2}dx$.

解 根据 D'Alembert 判别法可知 Euler-Poisson 积分收敛,引入参变量 t,使其成为 t 的函数.

令 $f(t)=\int_0^{+\infty}e^{-tx^2}dx$,设 $L[f(t)]=F(s)$,则

$$F(s)=\int_0^{+\infty}\left(\int_0^{+\infty}e^{-tx^2}dx\right)e^{-st}dt$$

$$=\int_0^{+\infty}\int_0^{+\infty}e^{-(x^2+s)t}dtdx$$

$$=\int_0^{+\infty}\frac{1}{s+x^2}\mathrm{d}x$$

$$=\frac{1}{\sqrt{s}}\int_0^{+\infty}\frac{1}{\left(\frac{x}{\sqrt{s}}\right)^2+1}\mathrm{d}\left(\frac{x}{\sqrt{s}}\right)$$

$$=\frac{\pi}{2}\cdot\frac{1}{\sqrt{s}}$$

根据 $F(s)$ 的反变换，由 $L[t^{-\frac{1}{2}}]=\frac{\sqrt{\pi}}{\sqrt{s}}$，可以推导出 $f(t)=\frac{1}{2}\cdot\sqrt{\frac{\pi}{t}}$，进而得出 $\int_0^{+\infty}\mathrm{e}^{-x^2}\mathrm{d}x=f(1)=\frac{\sqrt{\pi}}{2}$.

3　采用 Laplace 变换求解微分方程

利用 Laplace 变换求解微分方程的解题思路：

(1) 微分方程两端同时取 Laplace 变换，将常系数的微分方程转化为象函数的代数方程；

(2) 求解象函数满足的微分方程，得到象函数；

(3) 对象函数进行 Laplace 反变换，从而得到原方程的解.

例 6　求常微分方程 $y''(t)-3y'(t)+2y(t)=2\mathrm{e}^{3t}$，满足初值条件 $y(0)=2, y'(0)=3$ 的解.

解　设 $L\{y(t)\}=Y(s)$，在方程两端取 Laplace 变换可得

$$s^2Y(s)-sy(0)-y'(0)-3[sY(s)-y(0)]+2Y(s)=\frac{2}{s-3}$$

将初值条件代入，得到关于 $Y(s)$ 的代数方程为

$$Y(s)=\frac{2s^2-9s+11}{(s-3)(s-2)(s-1)} \tag{8}$$

利用待定系数法将式(8)分解为3个简单分式和的形式,即

$$Y(s)=\frac{2}{s-1}-\frac{1}{s-2}+\frac{5}{s-3}$$

再利用 Laplace 变换的逆变换,可以得到满足初始条件方程的解为

$$y(t)=2\mathrm{e}^{t}-\mathrm{e}^{2t}+5\mathrm{e}^{3t}$$

例 7 求解分段微分方程

$$y''(t)-5y'(t)+6y(t)=\begin{cases}0,0\leqslant t<1\\ \mathrm{e}^{-t},1\leqslant t\leqslant 2\\ 0,t>2\end{cases}$$

其中 $y(0)=0,y'(0)=0$.

解 对方程两边同时进行 Laplace 变换得

$$s^2Y(s)-5sY(s)+6Y(s)=\int_1^2\mathrm{e}^{-t}\cdot\mathrm{e}^{-st}\mathrm{d}t$$

整理得

$$Y(s)=\frac{-\mathrm{e}^{-2s}\mathrm{e}^{-2}}{(s+1)(s-2)(s-3)}+\frac{\mathrm{e}^{-s}\mathrm{e}^{-1}}{(s+1)(s-2)(s-3)} \tag{9}$$

对式(9)求 Laplace 反变换可以得到

$$y(t)=-\mathrm{e}^{-2}\left(\frac{1}{12}\mathrm{e}^{t-2}-\frac{1}{3}\mathrm{e}^{2(t-2)}+\frac{1}{4}\mathrm{e}^{3(t-2)}\right)\varepsilon(t-2)+\mathrm{e}^{-1}\left(\frac{1}{12}\mathrm{e}^{t-1}-\frac{1}{3}\mathrm{e}^{2(t-1)}+\frac{1}{4}\mathrm{e}^{3(t-1)}\right)\varepsilon(t-1)$$

例 8 求微分方程组

$$\begin{cases}y'(t)=2y(t)+x(t)\\ x'(t)=-y(t)+4x(t)\end{cases}$$

满足 $x(0)=0,y(0)=1$ 的解.

解 对方程两边同时进行 Laplace 变换得

$$\begin{cases} sY(s)-1=2Y(s)+X(s) \\ sX(s)=-Y(s)+4X(s) \end{cases} \tag{10}$$

对式(10)进行整理可得

$$X(s)=\frac{-1}{(s-3)^2}, Y(s)=\frac{s-4}{(s-3)^2}$$

取 $X(s)$, $Y(s)$ Laplace 反变换可得

$$x(t)=-te^{3t}, y(t)=(1-t)e^{3t}$$

当时在东北人民大学数学系任教的我国著名数学家徐利治教授早在1953年就发表了题为《富理、梅林变换的广义约当条件》[①]的论文. 徐教授指出:

Laplace 变换的复变反转公式的有效性可以从通常的 Jordan 条件里解放出来,使建立在 Saks 所定义的广义囿变的条件上. 关于这一点已在作者的文章 *On the inversion of the Laplace-Lebesgue integral* 中讨论过. 显然该文中的主要引理亦即作者另一文 *The representation of functions of bounded variation by singular integrals* 中所证结果的特殊情形中的一特例. 由于 *Laplace*, *Fourier*, *Mellin* 各种变换的反转性乃至于 *Poisson* 和的公式的成立均以 *Fourier* 的单积分公式或 *Dirichlet* 的核积分的解析为其共同的基础,因而本文的主要目的即在直接利用文 *On the inversion of the Laplace-Lebesgue integral*. 中关于 Dirichlet 核积

① 摘自《数学学报》,1953,3(2):142-146.

分的结果以减弱 Fourier 及 Mellin 变换的反转性的通常条件. 但在讨论 Mellin 变换时,我们还需要添证一个新引理.

我们的出发点便是下面的引理:

设 $\phi(t)$ 属于 $L(-\infty,+\infty)$ 且在可测点集 E 上是广义囿变的. 若 $t=t_0$ 为 E 的任一内点(interior point) 且 E 的测度均值函数 $\Delta(t_0,t)$ 在 t_0 附近为囿变,则当 t_0 为密度点(或疏散点) 时,我们有

$$\lim_{\lambda\to\infty}\frac{1}{\pi}\int_E \phi(t)\ \frac{\sin\lambda(t-t_0)}{t-t_0}\mathrm{d}t$$
$$=\frac{\phi(t_0+)+\phi(t_0-)}{2}(\text{或 }0)$$

此处 $\phi(t_0\pm)$ 表示在 E 集上当 $t\to t_0\pm$ 时 $\phi(t)$ 各处的极限.

值得指出,此引理亦可直接用 Saks 的一个引理及de la Vallée-Poussin 关于富理级数收敛测验法来证明. 但因结果为已知,故不重证. 作为本引理的一个简单应用,易推证下列的结果:

定理 1 设 $\phi(t)$ 属于 $L(-\infty,+\infty)$, 并设 $(-\infty,+\infty)$ 内有一间节 I 在其中存在一可测点集 E 使 $\phi(t)$ 在 E 及其补集 $\complement_E$(对 I 而言) 上均为广义囿变. 于是对于 E 的任一内点兼密度点 $t=x$ 且当 E 的测度均值函数 $\Delta(x,t)$ 在 x 邻近亦为囿变时,我们有

$$\frac{1}{2}[\phi(x+)+\phi(x-)]=\frac{1}{2\pi}\lim_{\lambda\to\infty}\int_{-\lambda}^{\lambda}\mathrm{e}^{-\mathrm{i}xu}\,\mathrm{d}u\int_{-\infty}^{+\infty}\phi(t)\mathrm{e}^{\mathrm{i}ut}\,\mathrm{d}t \tag{1}$$

显然上式亦即 $\phi(t)$ 的 Fourier 变换的反转公式. 通常所称的Jordan条件是指 $\phi(t)$ 在 $t=x$ 的邻域

（间节）内为囿变，而此处的条件是 $\phi(t)$ 无须在整个间节邻域内囿变. 故对于旧有定理的条件来说，是一种减弱. 关于上述结果的证明与文 *On the inversion of the Laplace-Lebesgue integral* 中定理 1 及 2 的推理相似，故在此从略.

在讨论 Mellin 变换的反转性的条件之前，先证明一个事实，即直线上的可测点集经过“对数式迁移”(exponential translation) 后，并不丧失密度点及测度均值函数在密度点附近的囿变性. 假定 E 为 $(0,+\infty)$ 内的一个可测点集，其中的点以 t 代表，则由对数式变换 $\xi=\log t(t>0)$ 所决定的 ξ 点集 G 即称为对数式迁移集. 因此我们所需要的引理便是：

若 t_0 为 E 的一个密度点且 E 的测度均值函数 $\Delta(t_0,t)$ 在 t_0 附近为囿变，则经过 $\xi=\log t$ 变换后相应的 $\xi_0=\log t_0$ 必为对数式迁移集 G 的密度点，且 G 的测度均值函数 $\overline{\Delta}(\xi_0,\xi)$ 在 ξ_0 点附近亦系囿变.

证 我们须引用 Kestelman 的 *Modern Theories of Integration* 一书（以下简称 MTI）中的若干现成结果. 设以 $X(t\mid E)$ 表示 E 的特征函数，即 $X(t\mid E)=1$（当 $t\in E$ 时）；$X(t\mid E)=0$（当 $t\notin E$ 时）. 又以 $E(t_0,t)$ 表示 E 与闭间节 $[t_0,t]$ 的交集，$m(E)$ 表示 E 的 Lebesgue 测度，$m_*(E)$，$m^*(E)$ 分别表示内外测度. 其余所用各种符号，其意义均与 MTI 上者同. 首先可证明

$$m(G(\xi_0,\xi))=\int_{t_0}^{t}\frac{1}{u}X(u\mid E)\mathrm{d}u \qquad (2)$$

今先证 $E(t_0,t)$ 的对数式迁移集 $G(\xi_0,\xi)$ 必为可测. 由 MTI 的定理 111 可知有开集序列 $\{O_t^{(n)}\}$ 存

在，使$O_t^{(n)} \supset O_t^{(n+1)} \supset E(t_0,t)(n=0,1,2,\cdots)$，且$m(\prod_{n=0}^{\infty} O_t^{(n)})=m(E(t_0,t))$. 相应地，如果$\overline{O}_\xi^{(n)}$表示$O_t^{(n)}$的迁移集，则得$\overline{O}_\xi^{(n)} \supset \overline{O}_\xi^{(n+1)} \supset G(\xi_0,\xi)$，$m(\prod_{n=0}^{\infty} \overline{O}_\xi^{(n)}) \geqslant m^*(G(\xi_0,\xi))$. 同理，如应用定理111的第二款，则有闭集序列$\{C_t^{(n)}\}$使$C_t^{(n)} \subset C_t^{(n+1)} \subset E(t_0,t)$，$m(\sum_{n=0}^{\infty}{}^* C_t^{(n)})=m(E(t_0,t))$. 在对数式变换下，相应地我们有$\sum_{n=0}^{\infty}{}^* \overline{C}_\xi^{(n)} \subset G(\xi_0,\xi)$，$m(\sum_{n=0}^{\infty}{}^* \overline{C}_\xi^{(n)}) \leqslant m_*(G(\xi_0,\xi))$. 由于$m(\prod_{n=0}^{\infty} O_t^{(n)})=m(\sum_{n=0}^{\infty}{}^* C_t^{(n)})$，$m(\prod_{n=0}^{\infty} O_t^{(n)}-\sum_{n=0}^{\infty}{}^* C_t^{(n)})=0$，易推证$m(\prod_{n=0}^{\infty} \overline{O}_\xi^{(n)}-\sum_{n=0}^{\infty}{}^* \overline{C}_\xi^{(n)})=0$(即零测度点集经对数式变换后还是零测度). 从而可知$m(\prod_{n=0}^{\infty} \overline{O}_\xi^{(n)})=m(\sum_{n=0}^{\infty}{}^* \overline{C}_\xi^{(n)})$. 因此$m^*(G(\xi_0,\xi))=m_*(G(\xi_0,\xi))$，亦即$G(\xi_0,\xi)$为可测的. 因而又知$X(v \mid G(\xi_0,\xi))$为可和的(summable).

又因$\log u$在$[t_0,t]$上系单调上升，且绝对连续(absolutely continuous)，故应用MTI上的定理275及277可得

$$m(G(\xi_0,\xi))=\int_{\log t_0}^{\log t} X(v \mid G(\xi_0,\xi))\mathrm{d}v(v=\log u)$$

$$=\int_{t_0}^{t} X(\log u \mid G(\log t_0,\log t)) \cdot$$

$$(\log u)'\mathrm{d}u$$

$$=\int_{t_0}^{t}\frac{1}{u}X(u\mid E)\mathrm{d}u \tag{3}$$

故式(2)已获证明.显然上述证法对于 $0<t<t_0$ 的情形还是一样适用.

这一步我们应考虑 G 在点 ξ_0 的测度均值函数

$$\overline{\Delta}(\xi_0,\xi)=\frac{m(G(\xi_0,\xi))}{\xi-\xi_0}$$

$$=\frac{1}{\log t-\log t_0}\int_{t_0}^{t}\frac{1}{u}X(u\mid E)\mathrm{d}u$$

$$t\neq t_0$$

由积分均值定理得

$$\overline{\Delta}(\xi_0,\xi)=\frac{\mu(t)}{\log t-\log t_0}$$

$$\int_{t_0}^{t}X(u\mid E)\mathrm{d}u\left(\frac{1}{t_0}\geqslant\mu(t)\geqslant\frac{1}{t}\right)$$

$$=\mu(t)\frac{t-t_0}{\log t-\log t_0}\Delta(t_0,t) \tag{4}$$

因 $\mu(t)\to\frac{1}{t_0},\Delta(t_0,t)\to 1(t\to t_0)$,故立得

$$\lim_{t\to t_0}\overline{\Delta}(\xi_0,\xi)=\frac{1}{t_0}\left(\frac{\mathrm{d}}{\mathrm{d}t}\log t\right)^{-1}_{t=t_0}=1$$

此即证明 ξ_0 确为 G 的密度点.

最后须证 $\overline{\Delta}(\xi_0,\xi)$ 为 ξ 的囿变函数.由于 $\mu(t)\geqslant\frac{1}{t}$,故当 $\delta>0$ 时

$$\mu(t)\int_{t_0}^{t+\delta}X(u\mid E)\mathrm{d}u$$

$$\geqslant\int_{t_0}^{t}\frac{1}{u}X(u\mid E)\mathrm{d}u+\int_{t}^{t+\delta}\frac{1}{u}X(u\mid E)\mathrm{d}u$$

$$=\mu(t+\delta)\int_{t}^{t+\delta}X(u\mid E)\mathrm{d}u$$

因此$\mu(t)\geqslant\mu(t+\delta)$，亦即$\mu(t)$为不增函数. 因此可见式(4)右端的各部分均系囿变函数，从而推知$\overline{\Delta}(\xi_0,\xi)$亦系囿变函数. 至此引理遂完全证明.

值得指出，以上的论证还可以加以扩充，以便推证比原引理更为一般的结论. 例如对数式变换可以改为$\xi=f(t)$，此处$f(t)$为一单调上升的连续函数.

接下来即可讨论 Mellin-Lebesgue 积分的反转，我们有：

定理 2 设E为$(0,+\infty)$内的可测点集，x为E的任一密度点(内点)，且E的测度均值函数$\Delta(x,t)$在x邻近为囿变. 于是对于任意一个函数$f(t)$，只要$t^{\gamma-1}f(t)$属于$L(0,+\infty)$且$f(t)$在靠近x的E的点集上为广义囿变，则下列的 Mellin-Lebesgue 积分式

$$F(s)=\int_{E}f(t)t^{s-1}\mathrm{d}t,s=\gamma+\mathrm{i}w \tag{5}$$

即可反转为

$$\frac{1}{2}(f(x+)+f(x-))=\frac{1}{2\pi\mathrm{i}}\lim_{\lambda\to\infty}\int_{\gamma-\mathrm{i}\lambda}^{\gamma+\mathrm{i}\lambda}F(s)x^{-s}\mathrm{d}s \tag{6}$$

证 令$E(0,+\infty)$的对数式迁移集为$G(-\infty,+\infty)$. 因此仿通常的办法，作对数式变换$\xi=\log t$(亦即$t=\mathrm{e}^{\xi}$)后，由(5)得

$$\begin{aligned}F(s)&=\int_{G(-\infty,\infty)}f(\mathrm{e}^{\xi})\mathrm{e}^{s\xi}\mathrm{d}\xi\\&=\int_{-\infty}^{\infty}f(\mathrm{e}^{\xi})X'X(\xi\mid G)\mathrm{e}^{\mathrm{i}\omega\xi}\mathrm{d}\xi\end{aligned} \tag{7}$$

因 $t^{\gamma-1}f(t)$ 属于 $L(0,+\infty)$，故经变换后可见 $f(\mathrm{e}^{\xi})\mathrm{e}^{\gamma\xi}X(\xi\mid G)$ 属于 $L(-\infty,+\infty)$. 由条件可知必存在一个 $t=x$ 点的邻域(a,b)，而 $f(t)$ 在交集 $E(a,b)$ 上为广义囿变. 相应地易看出函数 $f(\mathrm{e}^{\xi})\mathrm{e}^{\gamma\xi}$ 在交集$G(\log a,\log b)$上为广义囿变；又当ξ在G上时 $X(\xi\mid G)=1$. 故可见函数 $\Phi(\xi)=f(\mathrm{e}^{\xi})\mathrm{e}^{\gamma\xi}X(\xi\mid G)$ 在G集上为广义囿变. 根据引理可知$y=\log x$必为G的密度点，并且G的测度均值函数 $\overline{\Delta}(y,\xi)$ 在 $\xi=y$ 的附近亦系囿变. 因此，若当ξ在G的补集 $\complement_G$ 上时定义 $\Phi(\xi)=0$，则函数 $\Phi(\xi)$ 即完全满足定理 1 的条件. 从而利用公式(1) 可将式(7) 右端的富理变换在点$\xi=y$上加以反转

$$
\begin{aligned}
&\frac{1}{2}[f(\mathrm{e}^{y}+)\mathrm{e}^{\gamma y}+f(\mathrm{e}^{y}-)\mathrm{e}^{\gamma y}]\\
&=\frac{1}{2\pi}\lim_{\lambda\to\infty}\int_{-\lambda}^{\lambda}\mathrm{e}^{-\mathrm{i}yw}\,\mathrm{d}w\int_{-\infty}^{\infty}\Phi(\xi)\mathrm{e}^{\mathrm{i}w\xi}\,\mathrm{d}\xi\\
&=\frac{1}{2\pi}\lim_{\lambda\to\infty}\int_{-\lambda}^{\lambda}\mathrm{e}^{-\mathrm{i}yw}F(S)\,\mathrm{d}w\\
&=\frac{1}{2\pi\mathrm{i}}\lim_{\lambda\to\infty}\int_{\gamma-\mathrm{i}\lambda}^{\gamma+\mathrm{i}\lambda}F(S)\mathrm{e}^{-y(S-\lambda)}\,\mathrm{d}S
\end{aligned}
$$

自此式两端代回 $\mathrm{e}^{y}=x$，即得公式(6). 故定理 2 已证明.

显而易见，倘设 $E=(0,+\infty)$，则定理 2 即变为通常 Jordan 条件下的 Mellin 变换的反转定理. 又如仿照文 *On the inversion of the Laplace-Lebesgue integral* 中定理 2 的推理，则可知本文的定理 2 亦可略扩充为下述的形式：

定理 2′ 设 $t^{\gamma-1}f(t)$ 属于 $L(0,+\infty)$，并设$(0,+\infty)$ 内有一间节 I 在其中存在一可测点集 E 使

$f(t)$ 在 E 及其补集 $\complement_E$(对 I 言)上均系广义囿变. 于是对于 E 的任一密度点(内点)$t=x$ 且当 E 的测度均值函数 $\Delta(x,t)$ 在 x 邻近为囿变时,则下列的变换

$$F(s)=\int_0^{\infty} f(t)t^{s-1}\mathrm{d}t(s=\gamma+\mathrm{i}w) \tag{8}$$

常可反转成式(6).

注意在本文定理 1,2,2′ 中亦可将"测度均值函数 $\Delta(x,t)$ 为囿变"的条件一律换为比较特殊而简单的假设"x 为 E 的上升密度点".

最后我们指出,与 Fourier,Mellin 变换反转性的分析方式十分相似,Poisson 和的公式的论证亦可用广义囿变的条件来替代通常的 Jordan 条件,但其中须用及点集平移的办法,关于这些,此处不拟讨论.

1986 年安徽大学的沈克精教授在题为《有限亨克尔(Hankel)变换及其应用》[①] 的论文中把有限变换式的概念,推广到 Hankel 变换式的情形,建立相应的反演定理,并利用它求解关于具有轴对称性系统的边值问题. 这一类型的问题,通常是采用 Laplace 变换的方法求解,但此种方法很冗长,且往往需要计算复杂的围道积分. 而下文所要叙述的有限 Hankel 变换的方法却比较简捷,而且容易使用,它可避免用留数计算围道积分,如下:

① 摘自《安徽大学学报(自然科学版)》,1986(3):4-8.

1　有限 Hankel 变换及其反演定理

定义 1　$f(x)$ 在 $(0,1)$ 内的有限 Hankel 变换式定义为

$$F(p)=\int_0^1 xJ_n(px)f(x)\mathrm{d}x \tag{1}$$

其中 p 是方程

$$J_n(p)=0 \tag{2}$$

的正根.

$F(p)$ 称为 $f(x)$ 的有限 Hankel 变换，记为 $H[f(x)]=F(p)$，$f(x)$ 称为 $F(p)$ 的有限 Hankel 反变换，记为 $H^{-1}[F(p)]=f(x)$.

相仿地，仍用(1)定义有限 Hankel 变换，其中 p 是方程

$$pJ'_n(p)+hJ_n(p)=0 \tag{3}$$

的正根.

定义 2　$f(x)$ 在 (a,b) 内的有限 Hankel 变换式定义为

$$F(p)=\int_a^b xB_n(px)f(x)\mathrm{d}x, b>a \tag{4}$$

其中

$$B_n(px)=J_n(px)N_n(pa)-N_n(px)J_n(pa) \tag{5}$$

$N_n(px)$ 是第二类 n 阶 Bessel 函数，p 是方程

$$J_n(pb)N_n(pa)=N_n(pb)J_n(pa) \tag{6}$$

的正根.

从以上的定义，可建立相应的反演定理.

定理 1　设 $f(x)$ 在区间 $(0,1)$ 内，满足

Dirichlet 条件，而且是在(0,1) 内的有限 Hankel 变换式，由公式(1) 与(2) 所定义，则函数 $f(x)$ 在每一连续点处，均有

$$H^{-1}[F(p)]=f(x)=2\sum_{p}\frac{J_n(px)}{[J'_n(p)]^2}F(p) \quad (7)$$

其中 $\sum_{p}$ 是按方程(2) 的一切正根求和.

特别地，当 $n=0$ 时，注意到 $J'_0=-J_1(x)$，则得

$$H^{-1}[F(p)]=f(x)=2\sum_{p}\frac{J_0(px)}{J_1^2(p)}F(p) \quad (8)$$

定理 2　设 $f(x)$ 在区间 (0,1) 内，满足 Dirichlet 条件，而且在(0,1) 内的有限 Hankel 变换式，由公式(1) 与(3) 所定义，则函数 $f(x)$ 在每一连续点处，均有

$$H^{-1}[F(p)]=f(x)=2\sum_{p}\frac{J_n(px)}{J_n^2(p)}\frac{p^2F(p)}{h^2+p^2-n^2} \quad (9)$$

当 $n=0$ 时，得到

$$H^{-1}[F(p)]=f(x)=2\sum_{p}\frac{J_0(px)}{J_0^2(p)}\frac{p^2F(p)}{h^2+p^2} \quad (10)$$

定理 3　设 $f(x)$ 在区间 (a,b) 内，满足 Dirichlet 条件，而且(a,b) 内的有限 Hankel 变换式，由公式(4)(5) 及(6) 所定义，则函数 $f(x)$ 在每一连续点处，均有

$$H^{-1}[F(p)]=f(x)=\frac{\pi^2}{2}\sum_{p}B_n(px)\frac{p^2J_n^2(pb)}{J_n^2(pa)-J_n^2(pb)}F(p) \quad (11)$$

其中 $\sum\limits_{p}$ 是按方程(6) 的一切正根求和.

2　轴对称系统的有限 Hankel 变换

在极坐标系中,对于轴对称系统的定解问题与 θ 无关,泛定方程的一端具有如下形式

$$f(u)=\frac{1}{r}\frac{\partial}{\partial r}\left(r\frac{\partial u}{\partial r}\right)-\frac{n^2u}{r^2} \tag{12}$$

对(12) 施行有限 Hankel 变换,可得

$$\begin{aligned} H[f(u)] &= \int_0^1 J_n(pr)\frac{\partial}{\partial r}\left(r\frac{\partial u}{\partial r}\right)\mathrm{d}r-n^2\int_0^1\frac{u}{r}J_n(pr)\mathrm{d}r \\ &= \left[r\frac{\partial u}{\partial r}J_n(pr)\right]_0^1-p\int_0^1 rJ'_n(pr)\frac{\partial u}{\partial r}\mathrm{d}r- \\ &\quad n^2\int_0^1\frac{u}{r}J_n(pr)\mathrm{d}r \\ &= -p[urJ'_n(pr)]_0^1+p\int_0^1 u(J'_n(pr)+ \\ &\quad prJ''_n(pr)-n^2p^{-1}r^{-1}J_n(pr))\mathrm{d}r \end{aligned} \tag{13}$$

由于 $J_n(pr)$ 满足 Bessel 方程,即

$$prJ''_n(pr)+J'_n(pr)+(pr-n^2p^{-1}r^{-1})J_n(pr)=0$$

或

$$prJ''_n(pr)+J'_n(pr)-n^2p^{-1}r^{-1}J_n(pr)=-prJ_n(pr) \tag{14}$$

把(14) 代入(13) 得到

$$H[f(u)]=-pu_1J'_n(p)-p^2U \tag{15}$$

其中 $u_1=u\,|_{r=1}$,$U=H[u]$.

对于 p 是方程(3) 的根,可得到类似的结果,特别地,当 $n=0$ 时,得到

$$H[f(u)]=\int_0^1 \frac{\partial}{\partial r}\left(r\frac{\partial u}{\partial r}\right)J_0(pr)\mathrm{d}r$$

$$=J_0(p)\left[\frac{\partial u}{\partial r}+hu\right]_{r=1}-p^2U \quad (16)$$

对于由(4)(5)及(6)所定义的有限 Hankel 变换,得到的结果是

$$H[f(u)]=\int_a^b f(u)rB_n(pr)\mathrm{d}r$$

$$=\frac{2}{\pi}\left[u_b\frac{J_n(pa)}{J_n(pb)}-u_a\right]-p^2U \quad (17)$$

其中 $u_b=u|_{r=b}, u_a=u|_{r=a}$.

3 应用举例

例 1 径向热传导问题.

设半径为 1 的无界圆柱,初始温度为零,对于一切 $t>0$,侧面上保持恒温 u_0,求解柱内的温度分布规律.

解 所设条件归结为如下定解问题

$$\begin{cases}\dfrac{\partial^2 u}{\partial r^2}+\dfrac{1}{r}\dfrac{\partial u}{\partial r}=\dfrac{1}{k}\dfrac{\partial u}{\partial t}, 0\leqslant r<1, t>0 & (18)\\ u|_{r=1}=u_0, t>0 & (19)\end{cases}$$

$$\begin{cases}u|_{r=0}<+\infty, t>0 & (20)\\ u|_{t=0}=0, 0\leqslant r<1 & (21)\end{cases}$$

对(18)与(21)施行有限 Hankel 变换,利用(15)并注意到 $u_1=u_0$,得到

$$\begin{cases}\dfrac{\mathrm{d}u}{\mathrm{d}t}=-k[pu_0J_0'(p)+p^2U]\\ U|_{t=0}=0\end{cases}$$

解上述常微分方程,可得

$$U = u_0 p^{-1} J_0'(p)[\mathrm{e}^{-kp^2 t} - 1]$$

利用反演公式(8)得到

$$u = 2u_0 \sum_p (1 - \mathrm{e}^{-kp^2 t}) \frac{J_0(pr)}{pJ_1^2(p)}$$

其中$\sum\limits_p$是按方程$J_0(p)=0$的一切正根求和.

例 2 圆柱侧面具有对流的热流动问题.

设半径为 1 的无界圆柱,初始温度为 1,侧面上有与外边处于零温状态下的介质的对流的热交换,求解柱内的温度分布规律.

解 所设条件归结为下述定解问题

$$\begin{cases} \dfrac{\partial^2 u}{\partial r^2} + \dfrac{1}{r}\dfrac{\partial u}{\partial r} = \dfrac{1}{k}\dfrac{\partial u}{\partial t}, 0 \leqslant r < 1, t > 0 & (22) \\ \left(\dfrac{\partial u}{\partial r} + hu\right)\Big|_{r=1} = 0, t > 0 & (23) \\ u\,|_{t=0} < +\infty, t > 0 & (24) \\ u\,|_{t=0} = 1, 0 \leqslant r < 1 & (25) \end{cases}$$

其中u表示温度,k是材料的导温系数,h为常数.

对(22)(25)施行有限 Hankel 变换,并由边界条件(23)与(24)得到

$$\begin{cases} \dfrac{\mathrm{d}u}{\mathrm{d}t} = -kp^2 U \\ U\,|_{t=0} = \displaystyle\int_0^1 rJ_0(pr)\mathrm{d}r = \dfrac{J_1(p)}{p} \end{cases}$$

解上述常微分方程,得

$$U = p^{-1} J_1(p)_{\mathrm{e}} - kp^2 t$$

由反演公式(10)可得

$$u = 2\sum_p \mathrm{e}^{-kp^2 t} \frac{pJ_1(p)J_0(pr)}{h^2 + p^2 J^2(p)}$$

其中$\sum\limits_p$是按方程

$$pJ'_0(p)+hJ_0(p)=0$$

的一切正根求和.

例 3 两同心圆柱面间,粘滞流体的运动问题.

设有粘滞流体,包含在两个无界同心圆柱面之间,它们的半径分别是 a 与 b,内圆柱面固定不动,外柱面以固定的角速度 ω 突然开始旋转,求由此引起的流体运动规律.

解 设 v 是流体的运动速度,ρ 是运动粘滞系数,所设条件归结为下述定解问题

$$\begin{cases}\dfrac{\partial^2 v}{\partial r^2}+\dfrac{1}{r}\dfrac{\partial v}{\partial r}-\dfrac{v}{r^2}=\dfrac{1}{\rho}\dfrac{\partial v}{\partial t},a<r<b,t>0 & (26)\\ v\big|_{r=b}=\omega b,t>0 & (27)\\ v\big|_{r=a}=0,t>0 & (28)\\ v\big|_{t=0}=0,a<r<b & (29)\end{cases}$$

对(26) 与(29) 施行有限 Hankel 变换,并注意到边界条件(27) 与(28),得到

$$\begin{cases}\dfrac{\mathrm{d}\bar{v}}{\mathrm{d}t}=\rho\left[\dfrac{2}{\pi}\omega b\dfrac{J_1(pa)}{J_1(pb)}-p^2\bar{v}\right]\\ \bar{v}\big|_{t=0}=0\end{cases}$$

其中 $\bar{v}=H[v]$.

解上述常微分方程,得到

$$\bar{v}=\frac{2}{\pi}\frac{\omega b}{p^2}\frac{J_1(pa)}{J_1(pb)}(1-\mathrm{e}^{-\rho p^2 t})$$

利用反演公式(11) 得

$$v=\pi\omega b\sum_p\frac{1-\mathrm{e}^{-\rho p^2 t}}{J_1^2(pa)-J_1^2(pb)}J_1(pa)J_1(pb)B_1(pr) \tag{30}$$

(30) 中表示定常状态的项,可用如下的方法划分出来.

在(12) 中令 $n=1$,并记 $v=r-\frac{a^2}{r}$,于是

$$
\begin{aligned}
f(v) &= f\left(r-\frac{a^2}{r}\right) \\
&= \frac{\partial^2}{\partial r^2}\left(r-\frac{a^2}{r}\right)+\frac{1}{r}\frac{\partial}{\partial r}\left(r-\frac{a^2}{r}\right)-\frac{1}{r^2}\left(r-\frac{a^2}{r}\right) \\
&= 0
\end{aligned} \tag{31}
$$

$$
v_b=\left(r-\frac{a^2}{r}\right)\bigg|_{r=b}=\frac{b^2-a^2}{b} \tag{32}
$$

$$
v_a=\left(r-\frac{a^2}{r}\right)\bigg|_{r=a}=0 \tag{33}
$$

当 $n=1$ 时,把(31)(32)(33) 代入(17) 得

$$
\bar{v}=\frac{2}{\pi p^2}-\left(\frac{b^2-a^2}{b}\right)\frac{J_1(pa)}{J_1(pb)}
$$

即

$$
H\left[r-\frac{a^2}{r}\right]=\frac{2}{\pi p^2}\left(\frac{b^2-a^2}{b}\right)\frac{J_1(pa)}{J_1(pb)}
$$

由反演公式(11) 得

$$
\frac{r^2-a^2}{r}=\pi\left(\frac{b^2-a^2}{b}\right)\sum_p\frac{J_1(pa)J_1(pb)}{J_1^2(pa)-J_1^2(pb)}B_1(pr)
$$

于是(30) 可写成

$$
v=\frac{\omega b^2}{r}\left(\frac{r^2-a^2}{b^2-a^2}\right)\pi\omega b\sum_p\frac{J_1(pa)J_1(pb)}{J_1^2(pa)-J_1^2(pb)}B_1(pr)e^{-\rho p^2 t} \tag{34}
$$

其中 $\sum_p$ 是按方程 $B_1(pb)=0$ 的一切正根求和.(34) 的右端第一项表示定常状态.

本书是我们数学工作室庞大的原版引进计划中的一部.我们曾千百次与我们的读者相遇,不过我们从一开始就明白,让他们愿意向我们交付时间和信赖的,不是纸张、电磁

波、光信号等任何一种介质.连接我们彼此的是我们对这个世界共同的好奇心,对数学和物理学之美的探求,以及由此而生的对独立思考的执着,对人类智力成果的欣赏.

刘培杰

2022 年 12 月 18 日

于哈工大

刘培杰数学工作室

已出版(即将出版)图书目录——原版影印

书　　名	出版时间	定　价	编号
数学物理大百科全书.第1卷(英文)	2016—01	418.00	508
数学物理大百科全书.第2卷(英文)	2016—01	408.00	509
数学物理大百科全书.第3卷(英文)	2016—01	396.00	510
数学物理大百科全书.第4卷(英文)	2016—01	408.00	511
数学物理大百科全书.第5卷(英文)	2016—01	368.00	512

书　　名	出版时间	定　价	编号
zeta函数,q-zeta函数,相伴级数与积分(英文)	2015—08	88.00	513
微分形式:理论与练习(英文)	2015—08	58.00	514
离散与微分包含的逼近和优化(英文)	2015—08	58.00	515
艾伦·图灵:他的工作与影响(英文)	2016—01	98.00	560
测度理论概率导论,第2版(英文)	2016—01	88.00	561
带有潜在故障恢复系统的半马尔柯夫模型控制(英文)	2016—01	98.00	562
数学分析原理(英文)	2016—01	88.00	563
随机偏微分方程的有效动力学(英文)	2016—01	88.00	564
图的谱半径(英文)	2016—01	58.00	565
量子机器学习中数据挖掘的量子计算方法(英文)	2016—01	98.00	566
量子物理的非常规方法(英文)	2016—01	118.00	567
运输过程的统一非局部理论:广义波尔兹曼物理动力学,第2版(英文)	2016—01	198.00	568
量子力学与经典力学之间的联系在原子、分子及电动力学系统建模中的应用(英文)	2016—01	58.00	569

书　　名	出版时间	定　价	编号
算术域(英文)	2018—01	158.00	821
高等数学竞赛:1962—1991年的米洛克斯·史怀哲竞赛(英文)	2018—01	128.00	822
用数学奥林匹克精神解决数论问题(英文)	2018—01	108.00	823
代数几何(德文)	2018—04	68.00	824
丢番图逼近论(英文)	2018—01	78.00	825
代数几何学基础教程(英文)	2018—01	98.00	826
解析数论入门课程(英文)	2018—01	78.00	827
数论中的丢番图问题(英文)	2018—01	78.00	829
数论(梦幻之旅):第五届中日数论研讨会演讲集(英文)	2018—01	68.00	830
数论新应用(英文)	2018—01	68.00	831
数论(英文)	2018—01	78.00	832

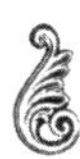

刘培杰数学工作室
已出版(即将出版)图书目录——原版影印

书　　名	出版时间	定　价	编号
湍流十讲(英文)	2018－04	108.00	886
无穷维李代数:第3版(英文)	2018－04	98.00	887
等值、不变量和对称性(英文)	2018－04	78.00	888
解析数论(英文)	2018－09	78.00	889
《数学原理》的演化:伯特兰·罗素撰写第二版时的手稿与笔记(英文)	2018－04	108.00	890
哈密尔顿数学论文集(第4卷):几何学、分析学、天文学、概率和有限差分等(英文)	2019－05	108.00	891
偏微分方程全局吸引子的特性(英文)	2018－09	108.00	979
整函数与下调和函数(英文)	2018－09	118.00	980
幂等分析(英文)	2018－09	118.00	981
李群,离散子群与不变量理论(英文)	2018－09	108.00	982
动力系统与统计力学(英文)	2018－09	118.00	983
表示论与动力系统(英文)	2018－09	118.00	984
分析学练习.第1部分(英文)	2021－01	88.00	1247
分析学练习.第2部分,非线性分析(英文)	2021－01	88.00	1248
初级统计学:循序渐进的方法:第10版(英文)	2019－05	68.00	1067
工程师与科学家微分方程用书:第4版(英文)	2019－07	58.00	1068
大学代数与三角学(英文)	2019－06	78.00	1069
培养数学能力的途径(英文)	2019－07	38.00	1070
工程师与科学家统计学:第4版(英文)	2019－06	58.00	1071
贸易与经济中的应用统计学:第6版(英文)	2019－06	58.00	1072
傅立叶级数和边值问题:第8版(英文)	2019－05	48.00	1073
通往天文学的途径:第5版(英文)	2019－05	58.00	1074
拉马努金笔记.第1卷(英文)	2019－06	165.00	1078
拉马努金笔记.第2卷(英文)	2019－06	165.00	1079
拉马努金笔记.第3卷(英文)	2019－06	165.00	1080
拉马努金笔记.第4卷(英文)	2019－06	165.00	1081
拉马努金笔记.第5卷(英文)	2019－06	165.00	1082
拉马努金遗失笔记.第1卷(英文)	2019－06	109.00	1083
拉马努金遗失笔记.第2卷(英文)	2019－06	109.00	1084
拉马努金遗失笔记.第3卷(英文)	2019－06	109.00	1085
拉马努金遗失笔记.第4卷(英文)	2019－06	109.00	1086
数论:1976年纽约洛克菲勒大学数论会议记录(英文)	2020－06	68.00	1145
数论:卡本代尔1979:1979年在南伊利诺伊卡本代尔大学举行的数论会议记录(英文)	2020－06	78.00	1146
数论:诺德韦克豪特1983:1983年在诺德韦克豪特举行的Journees Arithmetiques数论大会会议记录(英文)	2020－06	68.00	1147
数论:1985－1988年在纽约城市大学研究生院和大学中心举办的研讨会(英文)	2020－06	68.00	1148

刘培杰数学工作室
已出版(即将出版)图书目录——原版影印

书　　名	出版时间	定　价	编号
数论:1987 年在乌尔姆举行的 Journees Arithmetiques 数论大会会议记录(英文)	2020－06	68.00	1149
数论:马德拉斯 1987:1987 年在马德拉斯安娜大学举行的国际拉马努金百年纪念大会会议记录(英文)	2020－06	68.00	1150
解析数论:1988 年在东京举行的日法研讨会会议记录(英文)	2020－06	68.00	1151
解析数论:2002 年在意大利切特拉罗举行的 C. I. M. E. 暑期班演讲集(英文)	2020－06	68.00	1152
量子世界中的蝴蝶:最迷人的量子分形故事(英文)	2020－06	118.00	1157
走进量子力学(英文)	2020－06	118.00	1158
计算物理学概论(英文)	2020－06	48.00	1159
物质,空间和时间的理论:量子理论(英文)	2020－10	48.00	1160
物质,空间和时间的理论:经典理论(英文)	2020－10	48.00	1161
量子场理论:解释世界的神秘背景(英文)	2020－07	38.00	1162
计算物理学概论(英文)	2020－06	48.00	1163
行星状星云(英文)	2020－10	38.00	1164
基本宇宙学:从亚里士多德的宇宙到大爆炸(英文)	2020－08	58.00	1165
数学磁流体力学(英文)	2020－07	58.00	1166
计算科学:第 1 卷,计算的科学(日文)	2020－07	88.00	1167
计算科学:第 2 卷,计算与宇宙(日文)	2020－07	88.00	1168
计算科学:第 3 卷,计算与物质(日文)	2020－07	88.00	1169
计算科学:第 4 卷,计算与生命(日文)	2020－07	88.00	1170
计算科学:第 5 卷,计算与地球环境(日文)	2020－07	88.00	1171
计算科学:第 6 卷,计算与社会(日文)	2020－07	88.00	1172
计算科学. 别卷,超级计算机(日文)	2020－07	88.00	1173
多复变函数论(日文)	2022－06	78.00	1518
复变函数入门(日文)	2022－06	78.00	1523
代数与数论:综合方法(英文)	2020－10	78.00	1185
复分析:现代函数理论第一课(英文)	2020－07	58.00	1186
斐波那契数列和卡特兰数:导论(英文)	2020－10	68.00	1187
组合推理:计数艺术介绍(英文)	2020－07	88.00	1188
二次互反律的傅里叶分析证明(英文)	2020－07	48.00	1189
旋瓦兹分布的希尔伯特变换与应用(英文)	2020－07	58.00	1190
泛函分析:巴拿赫空间理论入门(英文)	2020－07	48.00	1191
卡塔兰数入门(英文)	2019－05	68.00	1060
测度与积分(英文)	2019－04	68.00	1059
组合学手册. 第一卷(英文)	2020－06	128.00	1153
*－代数、局部紧群和巴拿赫 *－代数丛的表示. 第一卷,群和代数的基本表示理论(英文)	2020－05	148.00	1154
电磁理论(英文)	2020－08	48.00	1193
连续介质力学中的非线性问题(英文)	2020－09	78.00	1195
多变量数学入门(英文)	2021－05	68.00	1317
偏微分方程入门(英文)	2021－05	88.00	1318
若尔当典范性:理论与实践(英文)	2021－07	68.00	1366
伽罗瓦理论. 第 4 版(英文)	2021－08	88.00	1408

刘培杰数学工作室
已出版(即将出版)图书目录——原版影印

书　名	出版时间	定　价	编号
典型群,错排与素数(英文)	2020—11	58.00	1204
李代数的表示:通过 gln 进行介绍(英文)	2020—10	38.00	1205
实分析演讲集(英文)	2020—10	38.00	1206
现代分析及其应用的课程(英文)	2020—10	58.00	1207
运动中的抛射物数学(英文)	2020—10	38.00	1208
2—纽结与它们的群(英文)	2020—10	38.00	1209
概率,策略和选择:博弈与选举中的数学(英文)	2020—11	58.00	1210
分析学引论(英文)	2020—11	58.00	1211
量子群:通往流代数的路径(英文)	2020—11	38.00	1212
集合论入门(英文)	2020—10	48.00	1213
酉反射群(英文)	2020—11	58.00	1214
探索数学:吸[illegible]证明方式(英文)	2020—11	58.00	1215
微分拓扑短期课程(英文)	2020—10	48.00	1216
抽象凸分析(英文)	2020—11	68.00	1222
费马大定理笔记(英文)	2021—03	48.00	1223
高斯与雅可比和(英文)	2021—03	78.00	1224
π 与算术几何平均:关于解析数论和计算复杂性的研究(英文)	2021—01	58.00	1225
复分析入门(英文)	2021—03	48.00	1226
爱德华·卢卡斯与素性测定(英文)	2021—03	78.00	1227
通往凸分析及其应用的简单路径(英文)	2021—01	68.00	1229
微分几何的各个方面.第一卷(英文)	2021—01	58.00	1230
微分几何的各个方面.第二卷(英文)	2020—12	58.00	1231
微分几何的各个方面.第三卷(英文)	2020—12	58.00	1232
沃克流形几何学(英文)	2020—11	58.00	1233
仿射和韦尔几何应用(英文)	2020—12	58.00	1234
双曲几何学的旋转向量空间方法(英文)	2021—02	58.00	1235
积分:分析学的关键(英文)	2020—12	48.00	1236
为有天分的新生准备的分析学基础教材(英文)	2020—11	48.00	1237
数学不等式.第一卷.对称多项式不等式(英文)	2021—03	108.00	1273
数学不等式.第二卷.对称有理不等式与对称无理不等式(英文)	2021—03	108.00	1274
数学不等式.第三卷.循环不等式与非循环不等式(英文)	2021—03	108.00	1275
数学不等式.第四卷.Jensen 不等式的扩展与加细(英文)	2021—03	108.00	1276
数学不等式.第五卷.创建不等式与解不等式的其他方法(英文)	2021—04	108.00	1277

刘培杰数学工作室
已出版(即将出版)图书目录——原版影印

书　　名	出版时间	定　价	编号
冯·诺依曼代数中的谱位移函数:半有限冯·诺依曼代数中的谱位移函数与谱流(英文)	2021－06	98.00	1308
链接结构:关于嵌入完全图的直线中链接单形的组合结构(英文)	2021－05	58.00	1309
代数几何方法.第1卷(英文)	2021－06	68.00	1310
代数几何方法.第2卷(英文)	2021－06	68.00	1311
代数几何方法.第3卷(英文)	2021－06	58.00	1312

书　　名	出版时间	定　价	编号
代数、生物信息和机器人技术的算法问题.第四卷,独立恒等式系统(俄文)	2020－08	118.00	1199
代数、生物信息和机器人技术的算法问题.第五卷,相对覆盖性和独立可拆分恒等式系统(俄文)	2020－08	118.00	1200
代数、生物信息和机器人技术的算法问题.第六卷,恒等式和准恒等式的相等 问题、可推导性和可实现性(俄文)	2020－08	128.00	1201
分数阶微积分的应用:非局部动态过程,分数阶导热系数(俄文)	2021－01	68.00	1241
泛函分析问题与练习:第2版(俄文)	2021－01	98.00	1242
集合论、数学逻辑和算法论问题:第5版(俄文)	2021－01	98.00	1243
微分几何和拓扑短期课程(俄文)	2021－01	98.00	1244
素数规律(俄文)	2021－01	88.00	1245
无穷边值问题解的递减:无界域中的拟线性椭圆和抛物方程(俄文)	2021－01	48.00	1246
微分几何讲义(俄文)	2020－12	98.00	1253
二次型和矩阵(俄文)	2021－01	98.00	1255
积分和级数.第2卷,特殊函数(俄文)	2021－01	168.00	1258
积分和级数.第3卷,特殊函数补充:第2版(俄文)	2021－01	178.00	1264
几何图上的微分方程(俄文)	2021－01	138.00	1259
数论教程:第2版(俄文)	2021－01	98.00	1260
非阿基米德分析及其应用(俄文)	2021－03	98.00	1261
古典群和量子群的压缩(俄文)	2021－03	98.00	1263
数学分析习题集.第3卷,多元函数:第3版(俄文)	2021－03	98.00	1266
数学习题:乌拉尔国立大学数学力学系大学生奥林匹克(俄文)	2021－03	98.00	1267
柯西定理和微分方程的特解(俄文)	2021－03	98.00	1268
组合极值问题及其应用:第3版(俄文)	2021－03	98.00	1269
数学词典(俄文)	2021－01	98.00	1271
确定性混沌分析模型(俄文)	2021－06	168.00	1307
精选初等数学习题和定理.立体几何.第3版(俄文)	2021－03	68.00	1316
微分几何习题:第3版(俄文)	2021－05	98.00	1336
精选初等数学习题和定理.平面几何.第4版(俄文)	2021－05	68.00	1335
曲面理论在欧氏空间 En 中的直接表示(俄文)	2022－01	68.00	1444
维纳—霍普夫离散算子和托普利兹算子:某些可数赋范空间中的诺特性和可逆性(俄文)	2022－03	108.00	1496
Maple 中的数论:数论中的计算机计算(俄文)	2022－03	88.00	1497
贝尔曼和克努特问题及其概括:加法运算的复杂性(俄文)	2022－03	138.00	1498

刘培杰数学工作室
已出版(即将出版)图书目录——原版影印

书 名	出版时间	定 价	编号
复分析:共形映射(俄文)	2022—07	48.00	1542
微积分代数样条和多项式及其在数值方法中的应用(俄文)	2022—08	128.00	1543
蒙特卡罗方法中的随机过程和场模型:算法和应用(俄文)	2022—08	88.00	1544
线性椭圆型方程组:论二阶椭圆型方程的迪利克雷问题(俄文)	2022—08	98.00	1561
动态系统解的增长特性:估值、稳定性、应用(俄文)	2022—08	118.00	1565
群的自由积分解:建立和应用(俄文)	2022—08	78.00	1570
混合方程和偏差自变数方程问题:解的存在和唯一性(俄文)	2023—01	78.00	1582
拟度量空间分析:存在和逼近定理(俄文)	2023—01	108.00	1583
二维和三维流形上函数的拓扑性质:函数的拓扑分类(俄文)	2023—03	68.00	1584
齐次马尔科夫过程建模的矩阵方法:此类方法能够用于不同目上的的复杂系统研究、设计和完善(俄文)	2023—03	68.00	1594

书 名	出版时间	定 价	编号
狭义相对论与广义相对论:时空与引力导论(英文)	2021—07	88.00	1319
束流物理学和粒子加速器的实践介绍:第 2 版(英文)	2021—07	88.00	1320
凝聚态物理中的拓扑和微分几何简介(英文)	2021—05	88.00	1321
混沌映射:动力学、分形学和快速涨落(英文)	2021—05	128.00	1322
广义相对论:黑洞、引力波和宇宙学介绍(英文)	2021—06	68.00	1323
现代分析电磁均质化(英文)	2021—06	68.00	1324
为科学家提供的基本流体动力学(英文)	2021—06	88.00	1325
视觉天文学:理解夜空的指南(英文)	2021—06	68.00	1326
物理学中的计算方法(英文)	2021—06	68.00	1327
单星的结构与演化:导论(英文)	2021—06	108.00	1328
超越居里:1903 年至 1963 年物理界四位女性及其著名发现(英文)	2021—06	68.00	1329
范德瓦尔斯流体热力学的进展(英文)	2021—06	68.00	1330
先进的托卡马克稳定性理论(英文)	2021—06	88.00	1331
经典场论导论:基本相互作用的过程(英文)	2021—07	88.00	1332
光致电离量子动力学方法原理(英文)	2021—07	108.00	1333
经典域论和应力:能量张量(英文)	2021—05	88.00	1334
非线性太赫兹光谱的概念与应用(英文)	2021—06	68.00	1337
电磁学中的无穷空间并矢格林函数(英文)	2021—06	88.00	1338
物理科学基础数学. 第 1 卷,齐次边值问题、傅里叶方法和特殊函数(英文)	2021—07	108.00	1339
离散量子力学(英文)	2021—07	68.00	1340
核磁共振的物理学和数学(英文)	2021—07	108.00	1341
分子水平的静电学(英文)	2021—08	68.00	1342
非线性波:理论、计算机模拟、实验(英文)	2021—06	108.00	1343
石墨烯光学:经典问题的电解解决方案(英文)	2021—06	68.00	1344
超材料多元宇宙(英文)	2021—07	68.00	1345
银河系外的天体物理学(英文)	2021—07	68.00	1346
原子物理学(英文)	2021—07	68.00	1347
将光打结:将拓扑学应用于光学(英文)	2021—07	68.00	1348
电磁学:问题与解法(英文)	2021—07	88.00	1364
海浪的原理:介绍量子力学的技巧与应用(英文)	2021—07	108.00	1365
多孔介质中的流体:输运与相变(英文)	2021—07	68.00	1372
洛伦兹群的物理学(英文)	2021—08	68.00	1373
物理导论的数学方法和解决方法手册(英文)	2021—08	68.00	1374

刘培杰数学工作室
已出版(即将出版)图书目录——原版影印

书　　名	出版时间	定　价	编号
非线性波数学物理学入门(英文)	2021—08	88.00	1376
波:基本原理和动力学(英文)	2021—07	68.00	1377
光电子量子计量学.第1卷,基础(英文)	2021—07	88.00	1383
光电子量子计量学.第2卷,应用与进展(英文)	2021—07	68.00	1384
复杂流的格子玻尔兹曼建模的工程应用(英文)	2021—08	68.00	1393
电偶极矩挑战(英文)	2021—08	108.00	1394
电动力学:问题与解法(英文)	2021—09	68.00	1395
自由电子激光的经典理论(英文)	2021—08	68.00	1397
曼哈顿计划——核武器物理学简介(英文)	2021—09	68.00	1401
粒子物理学(英文)	2021—09	68.00	1402
引力场中的量子信息(英文)	2021—09	128.00	1403
器件物理学的基本经典力学(英文)	2021—09	68.00	1404
等离子体物理及其空间应用导论.第1卷,基本原理和初步过程(英文)	2021—09	68.00	1405
拓扑与超弦理论焦点问题(英文)	2021—07	58.00	1349
应用数学:理论、方法与实践(英文)	2021—07	78.00	1350
非线性特征值问题:牛顿型方法与非线性瑞利函数(英文)	2021—07	58.00	1351
广义膨胀和齐性:利用齐性构造齐次系统的李雅普诺夫函数和控制律(英文)	2021—06	48.00	1352
解析数论焦点问题(英文)	2021—07	58.00	1353
随机微分方程:动态系统方法(英文)	2021—07	58.00	1354
经典力学与微分几何(英文)	2021—07	58.00	1355
负定相交形式流形上的瞬子模空间几何(英文)	2021—07	68.00	1356
广义卡塔兰轨道分析:广义卡塔兰轨道计算数字的方法(英文)	2021—07	48.00	1367
洛伦兹方法的变分:二维与三维洛伦兹方法(英文)	2021—08	38.00	1378
几何、分析和数论精编(英文)	2021—08	68.00	1380
从一个新角度看数论:通过遗传方法引入现实的概念(英文)	2021—07	58.00	1387
动力系统:短期课程(英文)	2021—08	68.00	1382
几何路径:理论与实践(英文)	2021—08	48.00	1385
论天体力学中某些问题的不可积性(英文)	2021—07	88.00	1396
广义斐波那契数列及其性质(英文)	2021—08	38.00	1386
对称函数和麦克唐纳多项式:余代数结构与Kawanaka恒等式(英文)	2021—09	38.00	1400
杰弗里·英格拉姆·泰勒科学论文集:第1卷.固体力学(英文)	2021—05	78.00	1360
杰弗里·英格拉姆·泰勒科学论文集:第2卷.气象学、海洋学和湍流(英文)	2021—05	68.00	1361
杰弗里·英格拉姆·泰勒科学论文集:第3卷.空气动力学以及落弹数和爆炸的力学(英文)	2021—05	68.00	1362
杰弗里·英格拉姆·泰勒科学论文集:第4卷.有关流体力学(英文)	2021—05	58.00	1363

刘培杰数学工作室
已出版(即将出版)图书目录——原版影印

书　　名	出版时间	定　价	编号
非局域泛函演化方程:积分与分数阶(英文)	2021－08	48.00	1390
理论工作者的高等微分几何:纤维丛、射流流形和拉格朗日理论(英文)	2021－08	68.00	1391
半线性退化椭圆微分方程:局部定理与整体定理(英文)	2021－07	48.00	1392
非交换几何、规范理论和重整化:一般简介与非交换量子场论的重整化(英文)	2021－09	78.00	1406
数论论文集:拉普拉斯变换和带有数论系数的幂级数(俄文)	2021－09	48.00	1407
挠理论专题:相对极大值,单射与扩充模(英文)	2021－09	88.00	1410
强正则图与欧几里得若尔当代数:非通常关系中的启示(英文)	2021－10	48.00	1411
拉格朗日几何和哈密顿几何:力学的应用(英文)	2021－10	48.00	1412
时滞微分方程与差分方程的振动理论:二阶与三阶(英文)	2021－10	98.00	1417
卷积结构与几何函数理论:用以研究特定几何函数理论方向的分数阶微积分算子与卷积结构(英文)	2021－10	48.00	1418
经典数学物理的历史发展(英文)	2021－10	78.00	1419
扩展线性丢番图问题(英文)	2021－10	38.00	1420
一类混沌动力系统的分歧分析与控制:分歧分析与控制(英文)	2021－11	38.00	1421
伽利略空间和伪伽利略空间中一些特殊曲线的几何性质(英文)	2022－01	68.00	1422
一阶偏微分方程:哈密尔顿—雅可比理论(英文)	2021－11	48.00	1424
各向异性黎曼多面体的反问题:分段光滑的各向异性黎曼多面体反边界谱问题:唯一性(英文)	2021－11	38.00	1425
项目反应理论手册.第一卷,模型(英文)	2021－11	138.00	1431
项目反应理论手册.第二卷,统计工具(英文)	2021－11	118.00	1432
项目反应理论手册.第三卷,应用(英文)	2021－11	138.00	1433
二次无理数:经典数论入门(英文)	2022－05	138.00	1434
数,形与对称性:数论,几何和群论导论(英文)	2022－05	128.00	1435
有限域手册(英文)	2021－11	178.00	1436
计算数论(英文)	2021－11	148.00	1437
拟群与其表示简介(英文)	2021－11	88.00	1438
数论与密码学导论:第二版(英文)	2022－01	148.00	1423

刘培杰数学工作室
已出版(即将出版)图书目录——原版影印

书　　名	出版时间	定　价	编号
几何分析中的柯西变换与黎兹变换:解析调和容量和李普希兹调和容量、变化和振荡以及一致可求长性(英文)	2021—12	38.00	1465
近似不动点定理及其应用(英文)	2022—05	28.00	1466
局部域的相关内容解析:对局部域的扩展及其伽罗瓦群的研究(英文)	2022—01	38.00	1467
反问题的二进制恢复方法(英文)	2022—03	28.00	1468
对几何函数中某些类的各个方面的研究:复变量理论(英文)	2022—01	38.00	1469
覆盖、对应和非交换几何(英文)	2022—01	28.00	1470
最优控制理论中的随机线性调节器问题:随机最优线性调节器问题(英文)	2022—01	38.00	1473
正交分解法:涡流流体动力学应用的正交分解法(英文)	2022—01	38.00	1475
芬斯勒几何的某些问题(英文)	2022—03	38.00	1476
受限三体问题(英文)	2022—05	38.00	1477
利用马利亚万微积分进行 Greeks 的计算:连续过程、跳跃过程中的马利亚万微积分和金融领域中的 Greeks(英文)	2022—05	48.00	1478
经典分析和泛函分析的应用:分析学的应用(英文)	2022—03	38.00	1479
特殊芬斯勒空间的探究(英文)	2022—03	48.00	1480
某些图形的施泰纳距离的细谷多项式:细谷多项式与图的维纳指数(英文)	2022—05	38.00	1481
图论问题的遗传算法:在新鲜与模糊的环境中(英文)	2022—05	48.00	1482
多项式映射的渐近簇(英文)	2022—05	38.00	1483
一维系统中的混沌:符号动力学,映射序列,一致收敛和沙可夫斯基定理(英文)	2022—05	38.00	1509
多维边界层流动与传热分析:粘性流体流动的数学建模与分析(英文)	2022—05	38.00	1510
演绎理论物理学的原理:一种基于量子力学波函数的逐次置信估计的一般理论的提议(英文)	2022—05	38.00	1511
R^2 和 R^3 中的仿射弹性曲线:概念和方法(英文)	2022—08	38.00	1512
算术数列中除数函数的分布:基本内容、调查、方法、第二矩、新结果(英文)	2022—05	28.00	1513
抛物型狄拉克算子和薛定谔方程:不定常薛定谔方程的抛物型狄拉克算子及其应用(英文)	2022—07	28.00	1514
黎曼-希尔伯特问题与量子场论:可积重正化、戴森-施温格方程(英文)	2022—08	38.00	1515
代数结构和几何结构的形变理论(英文)	2022—08	48.00	1516
概率结构和模糊结构上的不动点:概率结构和直觉模糊度量空间的不动点定理(英文)	2022—08	38.00	1517

刘培杰数学工作室

已出版(即将出版)图书目录——原版影印

书　　名	出版时间	定　价	编号
反若尔当对:简单反若尔当对的自同构(英文)	2022－07	28.00	1533
对某些黎曼－芬斯勒空间变换的研究:芬斯勒几何中的某些变换(英文)	2022－07	38.00	1534
内诣零流形映射的尼尔森数的阿诺索夫关系(英文)	2023－01	38.00	1535
与广义积分变换有关的分数次演算:对分数次演算的研究(英文)	2023－01	48.00	1536
强子的芬斯勒几何和吕拉几何(宇宙学方面):强子结构的芬斯勒几何和吕拉几何(拓扑缺陷)(英文)	2022－08	38.00	1537
一种基于混沌的非线性最优化问题:作业调度问题(英文)	即将出版		1538
广义概率论发展前景:关于趣味数学与置信函数实际应用的一些原创观点(英文)	即将出版		1539
纽结与物理学:第二版(英文)	2022－09	118.00	1547
正交多项式和 q－级数的前沿(英文)	2022－09	98.00	1548
算子理论问题集(英文)	2022－09	108.00	1549
抽象代数:群、环与域的应用导论:第二版(英文)	即将出版		1550
菲尔兹奖得主演讲集:第三版(英文)	2023－01	138.00	1551
多元实函数教程(英文)	2022－09	118.00	1552
球面空间形式群的几何学:第二版(英文)	2022－09	98.00	1566
对称群的表示论(英文)	2023－01	98.00	1585
纽结理论:第二版(英文)	2023－01	88.00	1586
拟群理论的基础与应用(英文)	2023－01	88.00	1587
组合学:第二版(英文)	2023－01	98.00	1588
加性组合学:研究问题手册(英文)	2023－01	68.00	1589
扭曲、平铺与镶嵌:几何折纸中的数学方法(英文)	2023－01	98.00	1590
离散与计算几何手册:第三版(英文)	2023－01	248.00	1591
离散与组合数学手册:第二版(英文)	2023－01	248.00	1592
分析学教程. 第 1 卷,一元实变量函数的微积分分析学介绍(英文)	2023－01	118.00	1595
分析学教程. 第 2 卷,多元函数的微分和积分,向量微积分(英文)	2023－01	118.00	1596
分析学教程. 第 3 卷,测度与积分理论,复变量的复值函数(英文)	2023－01	118.00	1597
分析学教程. 第 4 卷,傅里叶分析,常微分方程,变分法(英文)	2023－01	118.00	1598

联系地址:哈尔滨市南岗区复华四道街 10 号　哈尔滨工业大学出版社刘培杰数学工作室

网　　址:http://lpj.hit.edu.cn/

邮　　编:150006

联系电话:0451－86281378　　13904613167

E-mail:lpj1378@163.com